Gödel Without (Too Many) Tears

Gödel Without (Too Many) Tears

Second edition

Peter Smith

LOGIC MATTERS

Published by Logic Matters, Cambridge

ISBN 978-1-91690634-1 Hardback
ISBN 978-1-91690635-8 Paperback

Visit logicmatters.net/igt for further resources related to the topic of this book.

Contents

Preface

Why this short book? After all, I have already written a rather long book, *An Introduction to Gödel's Theorems*, originally published by CUP, now freely downloadable. Surely that's more than enough to be going on with?

Ah, but there's the snag. It *is* more than enough. In the writing, as is the way with these things, that book grew far beyond the scope of the original notes on which it was based. And while I hope the result is still quite accessible if you are prepared to put in the required time and effort, there is – to be frank – a *lot* more material in the book than is really needed by those wanting a first encounter with the famous incompleteness theorems.

Quite a few readers might therefore appreciate a cut-down version of some of that material – an introduction to the *Introduction*, if you like. Hence *Gödel Without (Too Many) Tears*. There are occasional footnotes referring to sections of the longer book, indicating where topics are discussed further: but you don't have to chase up those references to get a more limited but still coherent story in this shorter version.

What background do I presuppose? What do you need to bring to the party? Very little. If you have a grasp of a modest amount of logic, and have the patience to follow some simple mathematical arguments, you should have little difficulty in following the exposition here. I have given proofs of most of the important theorems I state, especially if the proofs involve some neat ideas. But I have left a few proofs for enthusiasts to follow up elsewhere, when trekking through the details would be too distracting or has little intrinsic interest.

GWT started life as a set of notes written to accompany the last outings of a short lecture course given in Cambridge (which was also repeated at the University of Canterbury, NZ). The notes aimed to bridge the gap between my classroom talk'n'chalk which just highlighted the Really Big Ideas, and the much more detailed treatments of topics available in *IGT*. However, despite that intended role, I did try to make the notes reasonably stand-alone.

Those original notes were tied to the first edition of *IGT*, as published in 2007. A significantly improved second edition of the book, *IGT2*, was published in 2013, which prompted me to revise the notes. Then came the pandemic in 2020; rewriting the notes again and turning them into a book became occupational therapy to distract me a little from the world's manifold troubles. The result was the first edition of *GWT* in book form.

This new version corrects known errors in the first edition, somewhat expands the material on computability and recursive functions, and makes a lot of small stylistic improvements, enough revisions overall to make it more than just a corrected reprint. So although the changes aren't radical, let's count it as a new edition, *GWT2*.

Many thanks to Henning Makholm for comments on the original notes, and also to David Auerbach and David Makinson for more comments that helped shape the resulting book version. I should also thank Ben Selfridge for pointing out the most serious glitch in that version, and so prompting me to start putting together this second edition. Many others have at various stages kindly let me know about typos and more serious mistakes, or made helpful suggestions. But I should especially mention Sam Butchart, David Furcy and Rowsety Moid for particularly helpful comments and corrections, both when I was writing the first edition of *GWT* and then again as I worked on *GWT2*. I really am very grateful to everyone!

1 A very brief note on Kurt Gödel

By common agreement, Kurt Gödel (1906–1978) was the greatest logician of the twentieth century.

Born in what is now Brno and educated in Vienna, Gödel left Austria for the USA in 1940, and spent the rest of his life at the Institute for Advanced Study at Princeton.

Gödel's doctoral dissertation, written when he was 23, established the *completeness* theorem for the predicate calculus (showing for the first time that a standard proof system for first-order logic does indeed capture all the semantically valid inferences).

Later he would do immensely important work on set theory, as well as make seminal contributions to proof theory and to the philosophy of mathematics. He even wrote about models of General Relativity with 'closed timelike curves' (where, in some sense, time travel is possible). But always a perfectionist, he became a very reluctant publisher: some of his philosophically most interesting work is in the substantial volume of Unpublished Essays and Lectures in his *Collected Works*.

Gödel proved a lot of important results, then. But talk of 'Gödel's Theorems' typically refers to the two *incompleteness* theorems he presented in an epoch-making 1931 paper.[1] And it is these theorems, and more particularly the First Theorem, that this book is all about. (Yes, that's right: Gödel did prove a 'completeness theorem' and also 'incompleteness theorems'. I'll explain the difference very soon.)

The impact of the incompleteness theorems on foundational studies is hard to exaggerate. For a start, putting it crudely and a bit tendentiously, they sabotage the ambitions of two major programmes in the foundations of mathematics – logicism and Hilbert's Programme.

We'll say just a little about logicism in the next chapter, and something about Hilbert's Programme much later, when we get round to discussing the Second Theorem in Chapter 19. But you don't have to know anything about this background to find the two theorems intrinsically fascinating. And as we will see, the beautiful ideas underlying their proofs are surprisingly easy to understand.

So now read on . . .

---

[1] Reprinted and translated in the first volume of Gödel's *Collected Works*. It is a masterpiece of compression, best tackled *after* you have read a book like this one!

2 Incompleteness, the very idea

The title of Gödel's great 1931 paper translates as *'On formally undecidable propositions of Principia Mathematica and related systems I'*.

The 'I' here indicates that this was intended to be the first part of a two part paper, with Part II spelling out in detail the proof of the Second Theorem which is only very briefly indicated in Part I. But Part II was never written. We'll see why not in due course.

This title already gives us a number of things to explain. What's a 'formally undecidable proposition'? What is *Principia Mathematica*? Ok, you've probably heard of that triple-decker work by A. N. Whitehead and Bertrand Russell, more than a century old and now very little read except by historians of logic: but what is the project of that book? And what counts as a 'related system' – a system suitably related, that is, to the one in *Principia*? In fact, just what is meant by 'system' here?

2.1 Effective decidability, effective computability

Taking the last question first, a 'system' (in the relevant sense) is an effectively axiomatized formal theory.[1] But what does that mean? We need, for a start, to explain 'effectively'.

Here are a couple of initial definitions to work with:

Defn. 1 *A property P (defined over some domain of objects D) is* effectively decidable *iff*[2] *there's an algorithm (a finite set of instructions for a deterministic computation) for settling in a finite number of steps, for any object $o \in D$, whether o has property P.*

To put it another way, a property is effectively decidable just when there's a step-by-step mechanical routine for settling whether o has property P, such that a suitably programmed deterministic computer could in principle implement the routine (idealizing away from practical constraints of time, etc.).

Here are two elementary and very familiar examples from propositional logic. The property of being a tautology is effectively decidable (by a truth-table test).

---

[1] It will turn out that Gödel originally had in mind just a central subclass of systems in this wide sense; but let's not complicate the story yet.

[2] 'Iff' is of course the logician's abbreviation for 'if and only if'.

The same goes for the property of being the main connective of a sentence (use bracket counting).

Relatedly, and even more fundamentally, we say:

Defn. 2 *A function f (a total function, defined over some domain of objects D) is* effectively computable *iff there's an algorithm for computing in a finite number of steps, for any object $o \in D$, the value of $f(o)$.*

Again, a function is effectively computable if an idealized computer, abstracting from practical limitations, could compute its value for any appropriate input by a finite number of operations. And we can tie our two definitions together. We can associate a property P with its *characteristic function c_P*, where $c_P(o) = 1$ iff o has property P and $c_P(o) = 0$ otherwise (think of the values '1' and '0' as representing 'true' and 'false'). Then P is effectively decidable if and only if c_P is effectively computable.

How satisfactory are our definitions, though? To elucidate them, we appealed to the idea of what an idealized computer could in principle do by implementing some algorithmic procedure. But this idea plainly stands in need of further elaboration. It turns out, however, that the notion of an effective computation is very robust: what is algorithmically-computable-in-principle according to one sensible sharpened-up definition is exactly the same as what is algorithmically-computable-in-principle according to any other sensible sharpened-up definition (see §17.4). Of course, it's not at all obvious that this is how things are going to pan out. So for the moment you are going to have to take it on trust (sorry!) that our two informal definitions *can* call upon a determinate formal notion of algorithmic computability.

Still, our current rough-and-ready explanation of 'effectively decidable' will suffice for present purposes in this chapter. Other uses of 'effective' can likewise be glossed along the lines of 'algorithmically computable'.

2.2   The idea of an effectively axiomatized formal theory

I'll take the general idea of an axiomatized theory to be pretty familiar. But now we need to be more specific: our focus is going to be on theories which have

  (a) an effectively formalized language,
  (b) an effectively decidable set of axioms, and
  (c) an effectively formalized proof system.

(a)   An *interpreted formalized language L* has both a *syntax* and an intended *semantics*:

  (1) *L*'s syntactic rules fix which strings of symbols form terms, which form *wffs* (i.e. well-formed formulas), and in particular which strings of symbols form *sentences*, i.e. closed wffs with no unbound variables dangling free.

  (2) *L*'s semantic rules assign interpretations – i.e., they assign truth-conditions – to every sentence of the language.

3

The standard way of presenting the syntax of a formal language is by first specifying some finite[3] set of basic logical and non-logical symbols, and then giving rules for building up more complex expressions from these symbols. And the rules are usually such that there are effective algorithmic procedures for deciding e.g. whether a given string of symbols counts as a term, or a wff, or a wff with one free variable, or a sentence; and there will be an effective procedure too for recovering from a sentence its 'constructional history', tracing the unique way it can be syntactically built up from its ultimate symbolic constituents.

The standard way of presenting the semantics starts by assigning truth-relevant values to the basic non-logical expressions of the language (e.g. by fixing the references of names and the extensions of predicates), and also we fix the domain(s) of quantification. Then we give algorithmically applicable rules for working out the truth-conditions of sentences in terms of the unique way they are syntactically built up from their parts. (Do read that carefully. What we should be able to work out mechanically is what the sentence *says*. But it is of course one thing to work out the conditions under which a sentence is true, and – usually – something quite different to work out whether those conditions are met, i.e. work out whether the sentence actually *is true*!)

So let's wrap that up in summary form:

Defn. 3 *An interpreted language L is effectively formalized iff (i) it has a finite set of basic symbols; (ii) syntactic properties such as being a term of the language, being a wff, being a wff with one free variable, and being a sentence, are all effectively decidable, and the syntactic structure of any sentence is effectively determinable; and (iii) this syntactic structure together with L's semantic rules can be used to effectively determine the unique intended interpretation of any sentence by fixing its truth-conditions.*

Now, *why* do we want (ii) the syntactic properties of being a sentence, etc., to be effectively decidable?[4] Well, the very point of setting up a formal language is to put such issues as what is and what isn't a well-formed sentence beyond dispute; and the best way of doing this is to ensure that even a suitably pro-grammed computer could decide whether a given string of symbols is indeed a sentence of the language.

And *why* do we want (iii) the unique truth-conditions of a sentence to be effectively determinable? Because we don't want any ambiguities or disputes about interpretation either.

(b)   A theory is sometimes defined by logic texts to be just any old set of sentences. We are concerned, though, with the more structured notion of an

---

[3] "Finite? But might we not need an unlimited, potentially infinite, supply of variables, for example?" Certainly. But we can in familiar ways build up an infinite list of variables from finite resources, e.g. as in 'x, x′, x″, x‴, . . .'. So we lose no relevant generality in keeping our basic symbol-set finite.

[4] It is in fact common to talk just about what is 'decidable'. But here at the outset it is probably helpful if I keep adding 'effectively' for emphasis.

axiomatized theory. In this case, we pick out some bunch of sentences Σ as giving *axioms* for the theory T; we also give T some *proof system*, i.e. some deductive apparatus; and then T's *theorems* are the sentences that are derivable from axioms in Σ by using the deductive apparatus.

An *effectively* axiomatized theory requires an effectively formalized language (and we are only going to be really interested in interpreted theories, with interpreted languages). But what else? For a start, we require it to have an effectively decidable set of axioms, meaning that the property of being a T-axiom is decidable. Why? Because if we are in the business of pinning down a theory by axiomatizing it, then we will normally want to avoid any possible dispute about what counts as a legitimate starting point for a proof by ensuring that we can mechanically decide whether a given sentence really is one of the axioms.

(c)   Just laying down a bunch of axioms would be pretty idle if we can't deduce conclusions from them! An axiomatized theory T will, as we said, come equipped with a proof system, a set of logical rules for deriving further theorems from our initial axioms. But a proof system such that we couldn't routinely tell whether its rules are being followed wouldn't have much point for practical purposes. Hence it is natural to require that T's logic has an effectively formalized proof system. In other words, we want it to be effectively decidable whether a given array of wffs is a well-constructed derivation from the axioms according to the rules of the proof system. It doesn't matter whether the proof system is an axiomatic logic, a natural deduction system, a tree/tableau system, or a sequent calculus – so long as it is effectively checkable whether a candidate proof-array has the property of being properly constructed according to the rules of the proof system.

Careful again! To say that it must be effectively decidable whether a candidate T-proof of φ is a kosher proof is not, repeat *not*, to say that it must be effectively decidable whether φ actually *has* a T-proof. To stress the point: it is one thing to be able to effectively *check* that some proposed proof follows the rules; it is another thing to be able to effectively *decide in advance* whether there exists a proof waiting to be discovered. Looking ahead, we will see as early as Chapter 5 that any effectively axiomatized formal theory T containing a modicum of arithmetic will be such that, although (i) you can mechanically check a proposed proof of φ to see whether it *is* a proof, sadly (ii) there's no mechanical way of telling in advance, given an arbitrary φ, whether it is provable in T.

So, in summary of (a) to (c),

Defn. 4 *An* effectively axiomatized *formal theory T has an effectively formalized language L, a certain class of L-wffs are picked out as axioms where it is effectively decidable what's an axiom, and it has a proof system such that it is effectively decidable whether a given array of wffs is a derivation from the axioms according to the rules.*

From now on, when we talk about formal theories, we will be concerned with effectively axiomatized formal theories (unless we explicitly say otherwise).

2.3 A quick reminder about logical proof systems

Let's recall some more logical notation. Working in some given formal language, suppose Σ is a set of sentences and φ is a particular sentence, and suppose S is a logical proof system. Then:

Defn. 5 '$\Sigma \vdash_S \varphi$' *says that there is a formal derivation in the proof system S from sentences in Σ to the sentence φ as conclusion.*

'$\Sigma \vDash \varphi$' *says that Σ logically entails φ, i.e. any way of (re)interpreting the relevant language's non-logical vocabulary that makes all the sentences in Σ true makes φ true too.*[5]

So '$\vdash_S$' signifies deducibility in S, which is a *syntactically* defined relation (being a well-formed proof is a question of being of the right symbolic shape, which is determined by syntactic pattern-matching). By contrast, '$\vDash$' signifies a *semantically* defined relation.

Of course, we normally want a formal deduction to be truth-preserving; so we will want our proof system S to respect logical entailments, requiring that if $\Sigma \vdash_S \varphi$ then indeed $\Sigma \vDash \varphi$. In a word, we require an acceptable logical proof system to be *sound*.

We can't in general insist on the converse. Not every relation of logical entailment can be captured in a proof system S for which it is effectively decidable what counts as an S-proof. But suppose we are working in a classical first-order setting, so the relevant logical vocabulary comprises just the truth-functional propositional connectives, the identity predicate, plus the apparatus of quantification. In this central case, if Σ logically entails φ, then there will indeed be a formal deduction of φ from Σ in your favourite first-order logical system S: i.e. if $\Sigma \vDash \varphi$ then $\Sigma \vdash_S \varphi$. In a word, S will be a *complete* deductive proof system for first-order logic.

As noted before, the first proof that a particular Hilbert-style axiomatic deductive system for first-order logic is complete was given by Gödel in his 1923 doctoral thesis: hence *Gödel's completeness theorem*.

2.4 'Formally undecidable propositions' and negation incompleteness

(a) We will recycle the familiar notation, for application to a formal theory T:

Defn. 6 '$T \vdash \varphi$' *says that there is a formal derivation in T's proof system from T's axioms to the sentence φ as conclusion (in short, φ is a T-theorem).*

---

[5] "Hold on! – aren't we taking our formal languages to have built-in interpretations: now you seem to be ignoring that." Not at all. Recall from elementary logic that some premises *logically* entail a given conclusion iff the inference from the premises to the conclusion is necessarily truth-preserving *just in virtue of the way that connectives and quantifiers (and any other logical operators) feature in the relevant sentences*. That's why our official definition of logical entailment abstracts away from the given interpretations of the non-logical constituents of the sentences while keeping the meaning of the logical apparatus fixed, and why we generalize over all possible reinterpretations of the non-logical constituents.

Now, we are interested in what claims a theory T can settle, one way or the other. So, assuming '$\neg$' is T's negation sign, we say

Defn. 7 *If T is a theory, and φ is some sentence of the language of that theory, then T formally decides φ iff either $T \vdash \varphi$ or $T \vdash \neg\varphi$. Hence, a sentence φ is formally undecidable by T iff $T \nvdash \varphi$ and $T \nvdash \neg\varphi$.*

A related bit of terminology:

Defn. 8 *A theory T is negation complete iff it formally decides every closed wff of its language – i.e. for every sentence φ, $T \vdash \varphi$ or $T \vdash \neg\varphi$.*

So there are formally undecidable propositions in a theory T if and only if T isn't negation complete.

(b)   It might help to fix ideas, and to distinguish the two notions of completeness – semantic completeness for a system of logic, negation completeness for a theory – if we look at a toy example.

Suppose then that theory T is built in a propositional language with just three propositional atoms, p, q, r, plus the usual propositional connectives. We give T a standard propositional classical logic (pick your favourite flavour of system!). And assign T just a single non-logical axiom: (p $\wedge \neg$r).

Then, just by assumption, T has a *semantically-complete logic*, since standard propositional calculi are complete. Hence, for any wff φ of T's limited language, if $T \vDash \varphi$, i.e. if T tautologically entails φ, then $T \vdash \varphi$.

However, trivially, T is not a *negation-complete theory*. For example T can't decide whether q is true. And there are lots of other wffs φ for which both $T \nvdash \varphi$ and $T \nvdash \neg\varphi$.

(c)   Our toy example shows that it is very, very easy to construct negation-incomplete theories with formally undecidable propositions: just hobble your theory T by leaving out some key assumptions about the matter in hand!

On the other hand, suppose we are trying to fully pin down some body of truths (e.g. the truths of basic arithmetic) using a formal theory T. We fix on an interpreted formal language L apt for expressing such truths. Then we'd ideally like to lay down enough axioms framed in L to give us a theory T such that, for any L-sentence φ, if φ is true then $T \vdash \varphi$. So, making the classical assumption that either φ is true or $\neg\varphi$ is true, we'd very much like T to be such that for any φ, either $T \vdash \varphi$ or $T \vdash \neg\varphi$ (but, of course, not both).

In other words, it is very natural to aim for theories T which are indeed negation complete.

2.5   Seeking a negation-complete theory of arithmetic

The elementary arithmetic of addition and multiplication is child's play (literally!). Surely we should be able to wrap it up in a nice formal theory, aiming for negation completeness!

So let's first fix on a formal *language of basic arithmetic* designed to express elementary arithmetical propositions. We start by giving this language

  (i) a term '0' to denote zero; and
  (ii) a sign 'S' for the successor (i.e. 'next number') function.

This means that we can construct the sequence of terms '0', 'S0', 'SS0', 'SSS0', ... to denote the natural numbers 0, 1, 2, 3, ... These are our language's *standard numerals*, and by using a standard numeral our language can denote any particular natural number.

We will also give this language

  (iii) function signs for addition and multiplication, together with
  (iv) the usual first-order logical apparatus including the identity sign, where
  (v) the quantifiers are interpreted as running over the natural numbers.

(We aren't building in subtraction and division as primitives. But these are definable in obvious ways.)

Now, it is entirely plausible to suppose that, whether or not the answers are readily available to us, questions posed in this language of basic arithmetic have entirely determinate answers. Why? Well, start from the following two bits of data:

  (a) The fundamental zero-and-its-successors structure of the natural number series.
  (b) The nature of addition and multiplication as given by the school-room explanations.

By (a) we mean that zero is not a successor, every number has a successor, distinct numbers have distinct successors, and so the sequence of zero and its successors never circles round but marches on for ever: moreover there are no strays – i.e. every natural number is in that sequence starting from zero. By (b) we mean to cover such basic laws as that $m + n = n + m$; we will say more about such laws in due course. It is very plausible to suppose that facts of the kind (a) and (b) together should fix the truth-value of every sentence of the language of basic arithmetic – after all, what more could it take?

But (a) and (b) seem so very basic and straightforward. So we will surely expect to be able to set down some axioms which (a) characterize the number series, and (b) define addition and multiplication: in other words, we should surely be able to frame axioms which codify what we teach the kids. And then the thought that (a) and (b) fix the truths of basic arithmetic becomes the thought that our axioms capturing (a) and (b) should settle every such truth. In other words, if φ is a true sentence of the language of successor, addition, and multiplication, then φ is provable from our axioms (and if φ is a false sentence, then $\neg\varphi$ is provable).

In sum, whatever might be the case with fancier realms of mathematics, it is very natural to suppose that we should at least be able to set down a negation-complete (and effectively axiomatized) formal theory of basic arithmetic.

2.6 Logicism and *Principia*

Now let's bring *Principia* into the story.

It is natural to ask: what is the *status* of the axioms of a formal theory of basic arithmetic? For example, what is the status of the formalized version of a truth like 'every number has a unique successor'? That hardly looks like a mere empirical generalization (something that could in principle be empirically refuted).

I suppose you might be a Kantian who holds that the axioms encapsulate 'intuitions' in which we grasp the fundamental structure of the numbers and the nature of addition and multiplication, where these 'intuitions' are a special cognitive achievement in which we somehow represent to ourselves an abstract arithmetical world.

But talk of such intuitions is, to say the least, puzzling and problematic. So we could very well be tempted instead by Gottlob Frege's seemingly more straightforward view that the axioms of arithmetic are *analytic*, simply truths of logic-plus-definitions. The idea is appealing: after all, every beginning logic student learns how to regiment 'If there are two *F*s and two *G*s, and none of the *F*s are *G*s, then there are four things which are *F*-or-*G*' as a theorem of first-order logic. So the idea that arithmetical truths like '2 + 2 = 4' can be treated as some kind of truth of logic seems rather natural. Perhaps arithmetic more generally is in some sense just part of logic. On this view, we don't need Kantian 'intuitions' going beyond logic: the proposal is that logical reasoning from mere definitions is enough to get us the axioms of arithmetic, and more logic gives us the rest of the arithmetic truths from these axioms.

This Fregean proposal is standardly dubbed *logicism*. But evidently, if this is to be more than wishful thinking, we need a well-worked-out logical system within which to pursue a logicist derivation of arithmetic. Famously, and to his eternal credit, Frege gave us the first competent system of quantificational logic in his *Begriffsschrift* of 1879. But equally famously, his own later attempt to go on to be a logicist about basic arithmetic (in fact, for Frege, about significantly more than basic arithmetic) hit the rocks, because – as Russell showed – the full deductive proof system that he then used, going beyond core quantificational logic, is inconsistent in a pretty elementary way. Frege's full system is beset by Russell's Paradox.[6]

That disaster devastated Frege; but Russell himself was undaunted. Still gripped by logicist ambitions he wrote:

> All mathematics [yes! – *all* mathematics] deals exclusively with concepts definable in terms of a very small number of logical concepts, and ... all its propositions are deducible from a very small number of fundamental logical principles.

That's a huge promissory note in Russell's *The Principles of Mathematics* (1903).

---

[6]Roughly, Frege's full system implies that there is a set of all sets which are not members of themselves – but ask: does that set belong to itself?

And *Principia Mathematica* (three volumes though unfinished, 1910, 1912, 1913) is Russell's attempt with Whitehead to start making good on that promise.

The project of *Principia*, in the roughest summary, is to set down some logical axioms and definitions from which we can deduce, for a start, all the truths of basic arithmetic (so giving us a negation-complete theory at least of arithmetic). Famously, the authors eventually get to prove that $1 + 1 = 2$ at *110.643 (Volume II, page 86), accompanied by the wry comment, 'The above proposition is occasionally useful'.

So far so good! But can Russell and Whitehead, in principle, prove *every* truth of arithmetic?

2.7 Gödel's bombshell

Principia, frankly, is a bit of a mess – in terms of clarity and rigour, it's quite a step backwards from Frege's logical systems. There are technical complications, and not all *Principia*'s axioms are clearly 'logical' even in a stretched sense. In particular, there's an appeal to a brute-force *Axiom of Infinity* which in effect stipulates that there is an infinite number of objects. But we don't need to go into details because we can leave such worries aside – they pale into insignificance compared with the bombshell exploded by Gödel.

For Gödel's First Incompleteness Theorem sabotages not just the overall project of *Principia* but – as advertised in the title of his paper – shows that *any* similar attempt to pin down *all* the truths of basic arithmetic in a theory with nice properties like being effectively axiomatized is in fatal trouble. His First Theorem says – at a good enough approximation for now – that *nicely formalized theories containing enough arithmetic are always negation incomplete.* Given any nice effectively axiomatized formal theory T, there will be arithmetic truths that can't be proved in that particular theory.

Only a moment ago, it didn't seem at all ambitious to try to capture all the truths of basic arithmetic in a single (consistent, effectively axiomatized) theory. But attempts to do so – and in particular, attempts to do this in a way that would appeal to Frege and Russell's logicist instincts – must always fail. Which is a rather stunning result![7]

So how did Gödel prove his result? The next chapter explains more carefully what the theorem (in two versions) claims, and then in Chapter 4 we outline a Gödelian proof of one version.

---

[7]'Hold on! I've heard of 'neo-logicism' which has its enthusiastic advocates. How can that be so if Gödel showed that logicism is a dead duck?'

Well, we might still like the idea that some logical principles plus what are more-or-less definitions (in a language richer than that of first-order logic) together *semantically* entail all arithmetical truths – even if we can't capture the relevant semantic entailment relation in a single effectively axiomatized deductive system of logic. Then the resulting overall system of arithmetic won't count as a formal effectively axiomatizable theory; so Gödel's theorems won't straightforwardly apply. But all that is another story.

3 The First Theorem, two versions

3.1 Soundness, consistency, etc.

Let's read into the record two more, no doubt familiar, definitions:

Defn. 9 *A theory T is* sound *iff its axioms are true (on the interpretation built into T's language) and its proof system is truth-preserving, so all its theorems are true.*

Defn. 10 *A theory T is (syntactically)* consistent *iff there is no φ such that $T \vdash \varphi$ and $T \vdash \neg\varphi$, where '$\neg$' is T's negation operator.*

Of course, soundness implies consistency. And in a classical setting, if T is inconsistent, then $T \vdash \varphi$ for all φ. So another way of defining consistency is by saying that T is consistent iff for some φ, $T \nvdash \varphi$. We shouldn't need to delay over these ideas.

But we also need another (quite natural) definition to use in this chapter:

Defn. 11 *The formalized interpreted language L* contains the language of basic arithmetic *iff L has a term which denotes zero and function symbols for the successor, addition and multiplication functions defined over numbers – these can be either built-in as primitives or introduced by definition – and has the usual connectives, the identity predicate, and can also express quantifiers running over the natural numbers.*

An example might be the language of set theory, in which we can define zero, successor, addition and multiplication in standard ways, and express restricted quantifiers running over just zero and its successors.[1]

3.2 Two theorems distinguished

In his 1931 paper, Gödel proves (more or less) the following: [2]

---

[1] Is the system of numbers referred to in set theory really the genuine article or just a structurally equivalent surrogate? We are not going to tangle with *that* messy issue! When we talk of quantifying over numbers inside e.g. set theory, understand this to be a matter of quantifying over whatever it is we can take to play the role of natural numbers.

[2] As footnoted at the beginning of Chapter 2, Gödel's initial idea of a formalized theory was in fact a bit narrower than our notion of an effectively axiomatized theory.

Theorem 1 *Suppose T is an effectively axiomatized formal theory whose language contains the language of basic arithmetic. Then, if T is sound, there will be a true sentence G_T of basic arithmetic such that $T \nvdash G_T$ and $T \nvdash \neg G_T$, so T is negation incomplete.*

We will outline a pivotal part of Gödel's proof in the next chapter.

However this version of an incompleteness theorem *isn't* what is most commonly referred to as *the* First Theorem, nor is it the result that Gödel foregrounds in his 1931 paper. For note, Theorem 1 tells us what follows from a *semantic* assumption, namely the assumption that T is sound. And soundness is defined in terms of truth.

Now, post-Tarski, most of us aren't particularly scared of the notion of truth. To be sure, there are issues about how best to treat the notion formally, to preserve as many as possible of our pre-formal intuitions while e.g. blocking the Liar Paradox. But most of us don't regard the relevant notion of a sound theory as metaphysically loaded in an obscure and worrying way. However, Gödel was writing at a time when – for various reasons (think logical positivism!) – the very idea of truth-in-mathematics was under some suspicion. It was therefore *extremely* important to Gödel that he could show that we don't need to deploy any semantic notions to get an incompleteness result. So he goes on to demonstrate a result which we can put schematically like this (more or less):

Theorem 2 *Suppose T is an effectively axiomatized formal theory whose language contains the language of basic arithmetic. Then, if T is consistent and can prove a certain modest amount of arithmetic (and has an additional property that any sensible formalized arithmetic will share), there will be a sentence G_T of basic arithmetic such that $T \nvdash G_T$ and $T \nvdash \neg G_T$, so T is negation incomplete.*

Being consistent is a syntactic property; being able to formally prove enough arithmetic is another syntactic property; and the mysterious additional property which I haven't explained is syntactically defined too. So *this* version of the incompleteness theorem only makes syntactic assumptions.

Of course, we'll need to be a lot more explicit about the details in due course; but this indicates the general character of Gödel's result in the second version. Our 'can prove a certain modest amount of arithmetic' gestures at what it takes for a theory to be sufficiently related to *Principia*'s system for the theorem to apply (recall the title of the 1931 paper). But I'll not pause here to spell out just how much arithmetic that is: we'll eventually find that it is stunningly little.[3]

For now, then, the first key take-away message of this chapter is that the incompleteness theorem does come in two different flavours. There's a version making a *semantic* assumption (the relevant theory T needs to be expressively rich enough and sound), and there's a version making only *syntactic* assumptions (about what T can and can't derive from its axioms).

---

[3]Nor will I pause to explain that 'additional property' condition. We'll meet it in due course, but also eventually see how – by a cunning trick discovered by J. Barkley Rosser in 1936 – we can drop that extra condition.

3.3 Incompleteness and incompletability

Let's concentrate for now on the first, semantic, version of the First Theorem.

Suppose, then, that T is a sound theory which contains the language of basic arithmetic. Then, the claim is, we can find a true G_T such that $T \nvdash \mathsf{G}_T$ and $T \nvdash \neg\mathsf{G}_T$. Let's be really clear: this doesn't, repeat *doesn't*, at all say that G_T is 'absolutely unprovable', whatever that obscure phrase could mean. It just says that G_T and its negation are *unprovable-in-T*.

Ok, you might very reasonably ask, why don't we simply 'repair the gap' in T by adding the true sentence G_T as a new axiom?

Well, consider the theory $U = T + \mathsf{G}_T$ (to use an obvious notation). Then (i) U is still sound, since the old T-axioms are true by assumption, the added new axiom is true, and the theory's logic is still truth-preserving. (ii) U is still a properly formalized theory; adding a single specified axiom to T doesn't make it undecidable what is an axiom of the augmented theory. (iii) U's language still contains the language of basic arithmetic. So Theorem 1 still applies, and we can find a sentence G_U such that $U \nvdash \mathsf{G}_U$ and $U \nvdash \neg\mathsf{G}_U$. And since U is stronger than T we have, a fortiori, $T \nvdash \mathsf{G}_U$ and $T \nvdash \neg\mathsf{G}_U$. In other words, 'repairing the gap' in T by adding G_T as a new axiom leaves some other sentences that were undecidable in T *still* undecidable in the augmented theory.

And so it goes. Keep throwing more and more additional true axioms at T and our theory will remain negation incomplete, unless it stops being effectively axiomatized. So here's the second key take-away message of the chapter: when the conditions for Theorem 1 apply, then the theory T will not just be incomplete but in a good sense T will be *incompletable* (or, as is sometimes said, T will be *essentially incomplete*).[4] We'll see in due course that the same holds when the conditions for Theorem 2 apply.

So we should perhaps really talk of the First *Incompletability* Theorem.

3.4 The completeness and incompleteness theorems again

We have already emphasized in §2.4 the distinction we need, and we illustrated it then with a toy example. But experience suggests that it will do no harm at all to repeat the point!

Suppose T is a theory of arithmetic cast in a first-order language, and equipped with a standard first-order deductive apparatus S. Then for any φ, if T logically entails φ then $T \vdash_S \varphi$. That's Gödel's completeness theorem for S.

But T can only too easily be a negation-incomplete theory of arithmetic. Just miss out axioms for addition (say), and there can be lots of wffs φ (those involving addition) such that neither $T \vdash \varphi$ nor $T \vdash \neg\varphi$!

---

[4]Suppose we take a theory with *all* the true sentences of the language of basic arithmetic as axioms. Then yes, by brute force, we get a negation-complete theory! What Theorem 1 will then tell us is that this theory can't be an effectively axiomatized theory – meaning that we can't effectively decide what's an axiom, i.e. we can't effectively decide what's a true sentence of the language. We'll be soon returning to this theme.

Of course, that's a *very* boring way of being negation incomplete. And, as we said before, we might reasonably have expected that such incompleteness can always be repaired by judiciously adding in the missing axioms. What the First Incompleteness Theorem tells us, however, is that try as we might, every theory of arithmetic satisfying certain elementary and highly desirable conditions (even if it has a semantically complete logic) must *remain* negation incomplete as a theory.

4 Outlining a Gödelian proof

4.1 A notational convention

Before continuing, we should highlight a notational convention that we have already started using:

(1) Expressions in informal mathematics will be in ordinary serif font, with variables, function names etc. in *italics*. Examples:

$$2 + 1 = 3,\ n + m = m + n,\ S(x + y) = x + Sy, Prf(m, n).$$

(2) Particular expressions from formal systems – and abbreviations of them – will be in sans serif type. Examples:

$$\mathsf{SSS0,\ SS0 + S0 = SSS0,\ \exists x\, x = 0,\ \forall x \forall y (x + y = y + x), Prf(x, y).}$$

(3) Greek letters, like 'Σ' and 'φ', are schematic variables in the metalanguage (i.e. in the language in which we are talking about our formal systems). So, in our case, these are variables added to logicians' English, used e.g. in generalizing about wffs of our formal systems.

In what follows, there will be a great deal of to-and-fro between (1) statements of informal mathematics, (2) formal expressions and formal proofs, and (3) general claims about formal expressions and formal proofs. It is essential for you to be clear which is which, and our (not unusual) notational convention should help you keep track.

4.2 Formally expressing numerical properties, relations and functions

In the next few sections, then, we are going to prepare the ground for §§4.6 and 4.7 where we give an outline sketch of how Gödel proved Theorem 1 (or at least, proved a very close relation).

We start with a couple more definitions. Recall, we said a language which includes the language of basic arithmetic will have (either built-in or defined) symbols '0' for zero and 'S' for the successor function. Then the standard numerals in such a language are the expressions '0', 'S0', 'SS0', 'SSS0', ...

Let's introduce a handy notational device:

Defn. 12 *We will use '$\overline{n}$' to abbreviate the standard numeral denoting the natural number n.*

15

So 'n̄' abbreviates n occurrences of 'S' followed by '0'. Thus '5̄' abbreviates 'SSSSS0', which in a formal language with standard numerals denotes what '5' denotes in informal arithmetical language.

Assume now that we are dealing with a language L which includes the language of basic arithmetic and so has standard numerals. Then we will say:

Defn. 13 *The open wff $\varphi(x)$ of the language L expresses the numerical property P just when, for any n, $\varphi(\bar{n})$ is true iff n has property P.*

Similarly, the formal wff $\psi(x, y)$ expresses the numerical two-place relation R just when, for any m and n, $\psi(\bar{m}, \bar{n})$ is true iff m has relation R to n.

And the formal wff $\chi(x, y)$ expresses the numerical one-place function f just when, for any m and n, $\chi(\bar{m}, \bar{n})$ is true iff $f(m) = n$.

Hopefully, this should all seem entirely natural.[1] For a couple of simple examples, the wff $\exists y\, x = (y + y)$ expresses the property of being an even number. Why? Because $\exists y\, \bar{n} = (y + y)$ is true just in case n is the sum of some natural number with itself, i.e. is twice some number. Similarly, $y = x \times x$ expresses the function which squares a number, because $\bar{n} = \bar{m} \times \bar{m}$ is true just in case $m^2 = n$.

Note, as we have defined it, for a wff to express the property of being an even number is just for it to be true of the even numbers, i.e. just for the interpreted wff to have the right *extension*. Consider the open wffs $\exists y\, x = ((S0 + S0) \times y)$ and $\exists y(x = (y + y) \wedge S0 + S0 = SS0)$. These differ in intuitive sense, but again are satisfied by just the even numbers, so also count as expressing the property of being even.

The point holds more generally: expressing a property, relation or function in our sense is just a matter of having the right extension.

The generalization of our Defn. 13 to cover wffs expressing many-place relations and many-place functions is obvious: we needn't pause to spell it out.

4.3 Gödel numbers

And now for an absolutely pivotal new idea.

These days, we are entirely familiar with the fact that all kinds of data can be coded up using numbers. The idea was certainly not in such everyday currency in 1931. But even then, the following sort of definition should have looked quite unproblematic:

---

[1]Fine print. '$\varphi(x)$' indicates, of course, a wff with one or more occurrences of the variable 'x' free. But of course, the particular choice of free variable doesn't matter. '$\varphi(\bar{n})$' then represents the sentence which results from replacing all free occurrences of the variable 'x' in $\varphi(x)$ by the standard numeral for n. As you knew!

If you've been well brought up, you might very well prefer the symbolism '$\varphi(\xi)$', which uses a metavariable ξ to mark a gap, rather than use '$\varphi(x)$' where we are recruiting the free variable 'x' for place-holding duties. But we will mostly stick to our more usual, slightly casual, mathematical usage, for familarity's sake (even though Fregeans will sigh sadly).

And a word to the wise: if you know what 'clash of variables' means, you will also know how we can avoid it in some future contexts by relabelling variables if necessary – so we just won't fuss about that.

Defn. 14 *A* Gödel-numbering scheme *for a formal theory T is some effective way of coding expressions of T (and sequences of expressions of T) as natural numbers. Such a scheme provides an algorithm for sending an expression (or sequence of expressions) to a number; and it also provides an algorithm for undoing the coding, which sends a code number back to the unique expression (or sequence of expressions) that it codes.*

Relative to a choice of scheme, the code number for an expression (or a sequence of expressions) is its unique Gödel number.

For a toy example, suppose the expressions of our theory's language L are built up from just eight basic symbols. Associate those with the digits 1 to 8, and associate the comma that we might use to separate expressions in a sequence of expressions with the digit 9. Then a single L-expression, and also a sequence of L-expressions separated by commas, can be directly mapped to a sequence of digits, which can then be read as a single numeral in standard decimal notation, denoting a natural number. That mapping is the simplest of algorithms. And in reverse, undoing the coding is equally simple and mechanical – though if the string of digits expressing some number contains the digit '0', the algorithm won't output any result when we try to decode it: assume our algorithm handles such cases gracefully.

Later we will want to focus on what we call 'normal' Gödel-numbering schemes (see §11.3). But for the present, all we will be assuming is that we use an effective scheme: otherwise it is just a matter of convenience which we might adopt.

4.4 Three new numerical properties/relations

Defn. 15 *Take an effectively axiomatized formal theory T, and fix on a scheme for Gödel-numbering expressions and sequences of expressions from T's language. Then, relative to that numbering scheme, we can define the following properties/relations:*

   i. *$Wff_T(n)$ iff n is the Gödel number of a T-wff.*
   ii. *$Sent_T(n)$ iff n is the Gödel number of a T-sentence.*
   iii. *$Prf_T(m, n)$ iff m is the Gödel number of a T-proof of the T-sentence with code number n.*

So Wff_T, for example, is a numerical property which, so to speak, 'arithmetizes' the syntactic property of being a T-wff.

Now, these three may not be numerical properties/relations of the familiar kind. But they are perfectly well-defined. And, crucially, we can say more:

Theorem 3 *Suppose T is an effectively axiomatized formal theory, and suppose we are given a Gödel-numbering scheme. Then the corresponding numerical properties/relations $Wff_T, Sent_T, Prf_T$ are effectively decidable.*

Proof Consider Wff_T. The number n has this property if and only if (i) n decodes into a string of T-symbols (by an effective procedure which a computer could carry out), and (ii) that string of symbols is indeed a T-wff (which, since T has an effectively formalized language by assumption, a computer could decide). Hence it is effectively decidable whether $\mathit{Wff}_T(n)$.

The case of Sent_T is similar. And as for Prf_T, since T is an effectively axiomatized theory it is effectively decidable whether a supposed proof-array of the theory is the genuine article proving its purported conclusion. So it is effectively decidable whether the array, if any, which gets the code number m is actually a T-proof of a sentence coded by n. That is to say, it is effectively decidable whether $\mathit{Prf}_T(m, n)$.[2] ⊠

Of course, just *which* numerical relation Prf_T (for example) is will depend on the details of the theory T and on our choice of Gödel-numbering scheme. But the key point is that so long as T is an effectively axiomatized formal theory, and so long as our coding scheme is algorithm-driven too, it must be a decidable property.

4.5 T can express Prf_T

So far, so straightforward. Now things get more exciting. In this section and the next, we state two key results, which will prepare the ground for our skeleton proof of Theorem 1. For the moment, we will have to state the results without proof; later, we will see what it takes to establish them. But at this point, we just want to explain what these two key results claim.

The first is as follows:

Theorem 4 *Suppose T is an effectively axiomatized formal theory which includes the language of basic arithmetic, and suppose we have fixed on a Gödel-numbering scheme. Then T can express the corresponding numerical relation Prf_T using some arithmetical wff* $\mathsf{Prf}_T(\mathsf{x}, \mathsf{y})$.

In other words, there is a wff $\mathsf{Prf}_T(\mathsf{x}, \mathsf{y})$ in the language of basic arithmetic such that $\mathsf{Prf}_T(\overline{\mathsf{m}}, \overline{\mathsf{n}})$ is true if and only if m codes for a T-proof of the sentence with Gödel number n.

This result is *not* supposed to be obvious! So how can we prove its perhaps surprising claim?

We can take the low road. Take a particular T and trudge through the details of building a wff of basic arithmetic which indeed expresses the relation Prf_T. Then we generalize, by noting that the same strategies and tricks that we use in the chosen particular case will apply equally when dealing with other effectively axiomatized formal theories.

Or we can take the high road. We start off by showing that, quite generally, the language of basic arithmetic has the resources to express *any* decidable properties and relations. And then we apply our sweeping result to the instances we are interested in: for we've just seen that the numerical relation Prf_T is decidable

---

[2]Our end-of-proof symbol will be '⊠': we need the more usual '□' for other duties later.

when T is an effectively axiomatized formalized theory. We will explore a version of this option in Chapter 11.

With a predicate Prf_T available in the theory T to express the relation Prf_T, we can now add a further simple definition:

Defn. 16 *Put* $\mathsf{Prov}_T(\mathsf{x}) =_{\mathrm{def}} \exists \mathsf{z} \mathsf{Prf}_T(\mathsf{z}, \mathsf{x})$ *(where the quantifier, if necessary, is restricted to run over the natural numbers in the domain).*

Then $\mathsf{Prov}_T(\overline{\mathsf{n}})$, *i.e.* $\exists \mathsf{z} \mathsf{Prf}_T(\mathsf{z}, \overline{\mathsf{n}})$, *is true iff some number Gödel-numbers a T-proof of the sentence with Gödel-number n, i.e. is true just if the sentence with code number n is a T-theorem. So* Prov_T *is naturally called a* provability predicate.

4.6 Defining a Gödel sentence G_T

And now comes the key result we need for building our skeletal proof of the First Theorem. Still working with an effectively axiomatized formal theory T whose language includes the language of basic arithmetic, and with a Gödel-numbering scheme in place:

Theorem 5 *We can construct a* Gödel *sentence* G_T *for the theory T in the language of basic arithmetic with the following property:* G_T *is true if and only if* $\neg\mathsf{Prov}_T(\ulcorner G_T \urcorner)$ *is true – where* $\ulcorner G_T \urcorner$ *is the numeral for the code number for* G_T.[3]

Don't worry for the moment about how we construct G_T (it is not difficult). Just note at the stage what our theorem implies. By construction, we said, G_T is true on interpretation iff $\neg\mathsf{Prov}_T(\ulcorner G_T \urcorner)$ is also true, i.e. iff the wff whose code number is denoted by $\ulcorner G_T \urcorner$ is not a T-theorem, i.e. iff G_T is not a T-theorem. In short, our theorem tells us that we can find an arithmetical sentence G_T which is *true if and only if it isn't a T-theorem.*

Stretching a point, it is rather as if G_T 'says' *I am unprovable in T.* Of course, strictly speaking, G_T doesn't *really* say that! – G_T is just a fancy sentence in the language of basic arithmetic, so it is in fact just about *numbers*, and doesn't refer to any wff. More about this later, in §12.2. Still, stretching the point will help you to spot that we can now immediately prove ...

4.7 Incompleteness!

Here again is

Theorem 1 *Suppose T is an effectively axiomatized formal theory whose language contains the language of basic arithmetic. Then, if T is sound, there will be a true sentence G_T of basic arithmetic such that $T \nvdash G_T$ and $T \nvdash \neg G_T$, so T is negation incomplete.*

---

[3]The rationale for our notational choice of $\ulcorner G_T \urcorner$ for the numeral will emerge later, §11.4!

Proof Take G_T to be the Gödel sentence introduced in Theorem 5. Suppose T is sound and $T \vdash G_T$. Then G_T would be a theorem, and hence G_T – which is true iff it is not a T-theorem – would be false. So T would have a false theorem and hence T would not be sound, contrary to hypothesis. So $T \nvdash G_T$.

Hence G_T is not provable. Since it is true iff it is not provable, G_T is true after all. So $\neg G_T$ is false and T, being sound, can't prove that either. Therefore we also have $T \nvdash \neg G_T$.

So, in sum, T can't formally decide G_T one way or the other. T is negation incomplete. ⊠

This proof, *once we have constructed* G_T, is very straightforward. So the devil is in the details of the proofs of the preliminary results we labelled as Theorems 4 and 5. As promised, later chapters will dig down to the relevant details.

Gödel's proof of the syntactic version of the incompleteness theorem, i.e. Theorem 2, also uses the same construction of a Gödel sentence, but this time we trade in the semantic assumption that T is sound for syntactic assumptions about what T can and can't prove. Therefore we will need syntactic analogues of Theorems 4 and 5. Again more devilish detail. Again more about this in later chapters.

4.8 Gödel and the Liar

So the claim is that, in a suitable theory T and using some Gödel coding, we can construct an arithmetic sentence G_T which as good as says that it is itself *unprovable* in T; and then such a sentence can neither be proved nor refuted in T assuming that theory is sound.

But you might well be suspicious. After all, we know we fall into paradox if we try to construct a Liar sentence L which as good as says that it is itself *not true*. So why does the construction of the Liar sentence lead to *paradox*, while the construction of the Gödel sentence gives us a *theorem*?

Which is a very good question. You have exactly the right instincts in raising it. The coming chapters, however, aim to give you a convincing answer.

But we are touching here on the deep roots of the incompleteness theorem. Suppose T is an effectively axiomatized theory which can express enough arithmetic. Then, as we'll confirm later, T can express the property of being a provable T-sentence. However, as we will also confirm, T can't express the property of being a true T-sentence (if it could, then T would be beset by the Liar paradox). So the property of being a true T-sentence and the property of being a provable T-sentence must be different properties. Hence either there are true-but-unprovable-in-T sentences or there are false-but-provable-in-T sentences. Assuming that T is sound rules out the second option. So the truths of T's language outstrip T's theorems. Therefore T can't be negation complete. *That* might be said to be the Master Argument for incompleteness: see §16.5.

5 Undecidability and incompleteness

Gödel's First Incompleteness Theorem tells us, roughly, that a nice enough theory T will always be negation incomplete for basic arithmetic.

We noted in Chapter 3 that the Theorem comes in two flavours, depending on whether we cash out the idea of being 'nice enough' in terms of (i) the semantic idea of T's being a *sound theory which uses enough of the language of arithmetic*, or (ii) the syntactic idea of T's being a *consistent theory which proves enough arithmetic*. Then we saw in Chapter 4 that Gödel's own proofs, of either flavour, go via the idea of numerically coding up syntactic facts about what can be proved in T, and then constructing an arithmetical sentence that – in virtue of the coding – is true if and only if it is not provable (it is rather as if it says *I am not provable in T*).

As we remarked, the Gödelian construction – at least as so far described – might look a bit worrying, with its echoes of the Liar Paradox. It might therefore go some way towards calming the worry that an illegitimate trick is being pulled if we now give a somewhat different proof of incompleteness. This proof will explicitly introduce the idea of a *diagonalization argument*. As we will see later, a form of diagonalization underlies Gödel's own proof.

5.1 The effective enumerability of theorems

We start with another definition invoking the informal notion of an effective procedure:

Defn. 17 *A set Σ is* effectively enumerable *iff there is an algorithmic routine which outputs a list of members of the set $s_1, s_2, s_3, \ldots$, repetitions allowed, such that any member of Σ will eventually appear on the list if the routine is run for enough steps.*

And to illustrate, here's an easy but important theorem:

Theorem 6 *The theorems of an effectively axiomatized theory T are effectively enumerable.*

Proof For convenience, we can assume our theory T's proof system is a Frege/ Hilbert axiomatic logic, where proofs are just linear sequences of wffs. But it should be pretty obvious how to generalize the argument to other kinds of proof systems, where proof arrays are arranged differently, for instance as trees of some kind.

Recall, we stipulated (in Defns. 3, 4) that if T is a properly formalized theory, its formalized language L has a finite number of basic symbols. Now, we can evidently put those basic symbols in some kind of 'alphabetical order', and then start mechanically listing off all the possible strings of symbols in order – e.g. the one-symbol strings, followed by the finite number of two-symbol strings in 'dictionary' order, followed by the finite number of three-symbol strings in 'dictionary' order, followed by the four-symbol strings, etc., etc.

Now, as we go along, generating strings of symbols, it will be a mechanical matter to decide whether a particular string is in fact a sequence of one or more wffs. And if it is, it will be a mechanical matter to decide whether the sequence of wffs is a T-proof, i.e. to check whether each wff is either an axiom or follows from earlier wffs in the sequence by one of T's rules of inference. (That's all effectively decidable in a properly formalized theory, by Defns. 3, 4). If the sequence *is* a kosher well-constructed proof, finishing with a sentence φ, then add this wff φ to our list of T-theorems.

We can in this way start mechanically generating a list which must eventually contain any T-theorem (since any T-theorem is the last sentence of a proof, and we eventually find any proof). ⊠

5.2 Negation completeness and decidability

Here is another natural definition:

Defn. 18 *A theory T is* decidable *iff the property of being a theorem of T is an effectively decidable property – i.e. iff there is a mechanical procedure for determining, for any given sentence φ of T's language, whether $T \vdash \varphi$.*

A terminology check is in order: a theory T formally *decides* a particular sentence φ iff either $T \vdash \varphi$ or $T \vdash \neg\varphi$; a theory T is *decidable* iff for *any* sentence φ of its language we can effectively determine whether $T \vdash \varphi$. Two quite different notions then, depite the similar terminology: in practice, though, you shouldn't get confused.[1]

Theorem 7 *Any consistent, negation-complete, effectively axiomatized formal theory T is decidable.*

Proof We have just shown that we can effectively enumerate the theorems of an effectively axiomatized formal theory. And that enables us to decide of an arbitrary sentence φ of our consistent, negation-complete T whether it is indeed a T-theorem.

---

[1] To fix ideas, note that a theory can be decidable without deciding every wff. For example, the toy propositional theory T of §2.4 is decidable (as is familiar, because propositional logic is complete, a truth-table test can be used to effectively determine whether $T \vdash \varphi$ for any given wff φ of T's language). In particular, we can thereby show that $T \nvdash \mathsf{q}$ and $T \nvdash \neg\mathsf{q}$. Therefore T doesn't decide q, so T doesn't decide every wff.

Just start enumerating the T-theorems. Since T is negation complete, eventually either φ or $\neg\varphi$ turns up (and then you can stop!). If φ turns up, declare it to be a theorem. If $\neg\varphi$ turns up, then since T is consistent, we can declare that φ is *not* a theorem.

Hence, there *is* a dumbly mechanical 'wait and see' procedure for effectively deciding whether φ is a T-theorem, a procedure which (given our assumptions about T) is guaranteed to deliver a verdict in a finite number of steps.   ⊠

We are, of course, relying here on our generously relaxed notion of effective decidability-in-principle, where we aren't working under any practical time constraints or constraints on available memory etc. (so, to emphasize, 'effective' doesn't mean 'practically efficacious' or 'efficient'). We might have to twiddle our thumbs for an immense time before one of φ or $\neg\varphi$ turns up. Still, our 'wait and see' method is guaranteed in this case to produce a result in finite time, in an entirely mechanical way.

So this counts as an effectively computable decision procedure in the official generous sense used in Defn. 1.

5.3   Capturing numerical properties in a theory

Here's an equivalent way of rewriting part of an earlier definition:

Defn. 13   *A numerical property P is expressed by the open wff $\varphi(\mathsf{x})$ with one free variable in a language L which contains the language of basic arithmetic iff, for every n,*
   i. if n has the property P, then $\varphi(\overline{n})$ is true,
   ii. if n does not have the property P, then $\neg\varphi(\overline{n})$ is true.

We now want a companion definition. Assume that the language of T includes the language of basic arithmetic so can form the standard numerals. Then:

Defn. 19   *The theory T captures the numerical property P by the open wff $\varphi(\mathsf{x})$ iff, for any n,*
   i. if n has the property P, then $T \vdash \varphi(\overline{n})$,
   ii. if n does not have the property P, then $T \vdash \neg\varphi(\overline{n})$.

Note the contrast: what a theory can *express* depends on the richness of its language (the definition doesn't mention proofs or theorems); what a theory can *capture* – mnemonic: <u>ca</u>se-by-case <u>p</u>rove – depends on what theorems can be derived in the theory, so depends on the richness of the theory's axioms.[2]

Just as a theory can express two-place relations (say) as well as monadic properties, a theory can capture relations as well as properties. So (for future reference) we expand our definition in the obvious way like this:

---

[2]To be honest, 'represents' is *much* more commonly used than my 'captures', but I'll stick here to the slightly idiosyncratic but more memorable jargon adopted in my *IGT*. Terminology here is a mess: for example, some use 'numeralwise express' to mean (not our 'express' but) 'captures/represents'.

Defn. 19 *(continued)   The theory T captures the two-place numerical relation R by the open wff $\varphi(\mathsf{x}, \mathsf{y})$ iff, for any m, n,*
      i. if m has the relation R to n, then $T \vdash \varphi(\overline{m}, \overline{n})$,
     ii. if m does not have the relation R to n, then $T \vdash \neg\varphi(\overline{m}, \overline{n})$.

And there is similar definition for the idea of capturing functions. But for the moment, let's concentrate on the case of capturing properties.

Ideally, of course, we will want any competent theory of arithmetic not just to express but also to capture lots of numerical properties, i.e. to be able to prove particular numbers have or lack these properties. But what kinds of properties do we want to capture?

Suppose that P is some effectively decidable property of numbers, i.e. one for which there is a mechanical procedure for deciding, given a natural number n, whether n has property P or not (see Defn. 1 again). So we can, in principle, run the procedure to decide whether n has this property P. Now, when we construct a formal theory of the arithmetic of the natural numbers, we will surely want deductions inside our theory to be able to track, case by case, any mechanical calculation that we can already perform informally (we see some examples of this in the next chapter). We don't want going formal to *diminish* our ability to determine whether n has the decidable numerical property P. Formalization aims at regimenting what we can in principle already do: it isn't supposed to hobble our efforts. So while we might have some passing interest in more limited theories, we will ideally aim for a formal theory T which at least (i) is able to frame some open wff $\varphi(\mathsf{x})$ which expresses our decidable numerical property P, and (ii) is such that if n has property P, $T \vdash \varphi(\overline{n})$, and if n does not have property P, $T \vdash \neg\varphi(\overline{n})$.

In short, then, we will want T not only to be able to *express* the decidable numerical property P but also to be able to *capture* P in the sense of our definition.

Focusing on the syntactic side of this, let's say more generally:

Defn. 20 *A formal theory T is sufficiently strong iff it captures all effectively decidable numerical properties.*[3]

Then, in summary, it seems reasonable to want a formal theory of arithmetic to be sufficiently strong. When *we* can (or at least, given world enough and time, *could*) decide of any particular number whether it has a certain property, the *theory* should be able to do that too.

5.4   Sufficiently strong theories are undecidable

We now establish a lovely theorem (and do take its elegant proof by 'diagonalization' slowly – savour it!):

---

[3]It would be equally natural, of course, to require that the theory also capture all decidable relations and all computable functions – but for present purposes we don't need to add that.

Theorem 8 *No consistent, effectively axiomatized and sufficiently strong formal theory is decidable.*

Proof We suppose T is a consistent, effectively axiomatized, and sufficiently strong theory yet is also decidable, and now derive a contradiction.

Given T is sufficiently strong, it must have a supply of open wffs suitable for capturing numerical properties. And by Defn 3 it must be decidable what strings of symbols are suitable T-wffs with one variable occurring free. So, we can use the same idea as in the proof of Theorem 6 to start effectively enumerating such wffs, which can represent by

$$\varphi_0(\mathsf{x}), \varphi_1(\mathsf{x}), \varphi_2(\mathsf{x}), \varphi_3(\mathsf{x}), \ldots.$$

For we can just start churning out all the strings of symbols of T's language (by length and in 'alphabetical order'), and as we go along we mechanically select out the wffs with one free variable.

We can then introduce the following definition of the numerical property D:

(*) n has the property D if and only if $T \vdash \neg\varphi_n(\overline{\mathsf{n}})$.

That's a perfectly coherent stipulation. Of course, property D isn't presented in the familiar way in which we ordinarily present properties of numbers: but our definition tells us what has to be the case for n to have the property D, and that's all we will need.

Now for the key observation: our supposition that T is a decidable theory entails that D is an effectively decidable property of numbers.

Why? Because, given any number n, it will be a mechanical matter to start listing off the one-free-variable wffs until we get to the n-th one, $\varphi_n(\mathsf{x})$. Then it is a mechanical matter to form the numeral $\overline{\mathsf{n}}$, substitute it for the variable, and then prefix a negation sign. Now we just apply the supposed mechanical procedure for deciding whether a sentence is a T-theorem to test whether the resulting wff $\neg\varphi_n(\overline{\mathsf{n}})$ is a theorem. So, on our current assumptions, there is an algorithm for deciding whether n has the property D.

Since, by hypothesis, the theory T is sufficiently strong, it can capture all decidable numerical properties. Hence it follows, in particular, that D is capturable by some open wff. This wff must of course eventually occur somewhere in our list of the $\varphi(\mathsf{x})$. Let's suppose the d-th wff does the trick: that is to say, suppose property D is captured by $\varphi_d(\mathsf{x})$.

It is now entirely routine to get out a contradiction. For, just by the definition of capturing, to say that $\varphi_d(\mathsf{x})$ captures D means that for any n,

> if n has the property D, $T \vdash \varphi_d(\overline{\mathsf{n}})$,
> if n doesn't have the property D, $T \vdash \neg\varphi_d(\overline{\mathsf{n}})$.

So taking in particular the case $n = d$, we have

   i. if d has the property D, $T \vdash \varphi_d(\overline{\mathsf{d}})$,
   ii. if d doesn't have the property D, $T \vdash \neg\varphi_d(\overline{\mathsf{d}})$.

But note what our initial definition (*) of the property D implies for the particular case $n = d$:

iii. d has the property D if and only if $T \vdash \neg\varphi_d(\overline{\mathsf{d}})$.

From (ii) and (iii), it follows that whether d has property D or not, the wff $\neg\varphi_d(\overline{\mathsf{d}})$ is a theorem either way. So by (iii) again, d does have property D, hence by (i) the wff $\varphi_d(\overline{\mathsf{d}})$ must be a theorem too. So a wff and its negation are both theorems of T. Which contradicts our assumption that T is consistent.   ⊠

In sum, if T is an effectively axiomatized formal theory, we will be able to effectively decide whether a purported proof of a particular T-sentence φ is the genuine article: but if T is also consistent and sufficiently strong, there is no general effective way of deciding in advance whether φ has a proof or not.

5.5   A word about 'diagonalization'

Let's highlight the key construction here involved in defining the property D. For each n, we take the n-th wff $\varphi_n(\mathsf{x})$ in our list, and plug in the standard numeral for the place-index n (before taking the negation of the result). This sort of thing is called *diagonalization*. Why? Just consider the square array you get by writing

$$\varphi_0(\overline{0}) \quad \varphi_0(\overline{1}) \quad \varphi_0(\overline{2}) \quad \varphi_0(\overline{3}) \ \ldots$$
$$\varphi_1(\overline{0}) \quad \varphi_1(\overline{1}) \quad \varphi_1(\overline{2}) \quad \varphi_1(\overline{3}) \ \ldots$$
$$\varphi_2(\overline{0}) \quad \varphi_2(\overline{1}) \quad \varphi_2(\overline{2}) \quad \varphi_2(\overline{3}) \ \ldots$$
$$\varphi_3(\overline{0}) \quad \varphi_3(\overline{1}) \quad \varphi_3(\overline{2}) \quad \varphi_3(\overline{3}) \ \ldots$$
$$\ldots \qquad \ldots \qquad \ldots \qquad \ldots \quad \searrow$$

Evidently, the wffs of the form $\varphi_n(\overline{\mathsf{n}})$, including $\varphi_d(\overline{\mathsf{d}})$, lie down the diagonal through the array.

We'll be meeting other instances of this sort of diagonal construction. And it is a diagonalization of this general kind that is really at the heart of Gödel's incompleteness proof.[4] More about this in due course.

5.6   Incompleteness again!

So we have now shown:

Theorem 7 *Any consistent, negation-complete, effectively axiomatized formal theory T is decidable.*

Theorem 8 *No consistent, effectively axiomatized and sufficiently strong formal theory is decidable.*

We can therefore immediately deduce:

---

[4]The grandfather of all such uses of diagonalization is Cantor's diagonal argument to show a set can't be equinumerous with its powerset.

Theorem 9 *A consistent, effectively axiomatized, sufficiently strong, formal theory cannot be negation complete.*

Wonderful! A seemingly remarkable theorem, proved remarkably quickly (this time without having to simply assume some as-yet-unproved theorems along the way).[5]

Note, though, that – unlike Gödel's own proof strategy – Theorem 9 doesn't actually yield a specific undecidable sentence for a given theory T. And more importantly, the interest of the theorem depends on the still-informal notion of a 'sufficiently strong' theory being in good order. Have we perhaps just shown that looking for sufficient strength is, after all, an unreasonable demand?

Now, I wouldn't have written up the argument in this chapter if this notion of T's being sufficiently strong were intrinsically problematic. Still, we are left with a major task here: we will need to give a sharper account of what makes for an effectively decidable property in order to (i) clarify the notion of sufficient strength, while (ii) making it plausible that we really do want theories to be sufficiently strong in this clarified sense.

This can be done. However, supplying and defending the needed sharp account of the notion of effective decidability takes quite a bit of work (see Chapter 18). And in fact we don't need to do all that work in order to prove core versions of the First Incompleteness Theorem via Gödel's original method as partially sketched in Chapter 4. So, over the coming chapters, we are going to start by reverting to exploring something closer to Gödel's route to the incompleteness theorems.

---

[5]I learnt our argument for Theorem 9 as a student – so decades ago! – from lectures by Timothy Smiley.

6 Two weak arithmetics

So far we have talked rather abstractly of theories which 'can prove a certain modest amount of arithmetic' and about theories which are 'sufficiently strong'. But we haven't said anything about what such theories look like. It is obviously high time that we stopped operating at the level of abstraction of earlier chapters; we need to start getting down to details.

This chapter, then, introduces a couple of weak arithmetics ('arithmetics', that is to say, in the sense of 'theories of arithmetic'). We first meet Baby Arithmetic and then the important Robinson Arithmetic. You can by all means skip lightly over some of the more boring proof details here; but you do need to get a sense of how these two weaker formal theories work, in preparation for the next chapter where we introduce the much stronger Peano Arithmetic.

6.1 The language L_B

(a) First we describe *the language of baby arithmetic, L_B.*

Its symbols, with their built-in interpretations, are

0	constant, denoting zero
S, +, ×	function symbols for, respectively, the successor, addition and multiplication functions
=, ¬, →	the identity predicate, negation, and conditional
(,)	parentheses for use with +, × and →.

We could give our language the other propositional connectives if we like: but the crucial thing is that L_B lacks the apparatus of quantification.

We write the one-place successor function symbol in prefix position, so we can form the standard numerals 0, S0, SS0, SSS0, . . . (see §2.5). Recall, we use 'n̄' to represent the standard numeral SS . . . S0 with n occurrences of 'S'.

We will however write '+' and '×' as infix function symbols in the usual way – i.e. we write (S0 + SS0) rather than prefix the function sign as in +S0SS0. So we need the parentheses for scoping the function signs, to disambiguate S0 + SS0 × SSS0, e.g. as (S0 + (SS0 × SSS0)). For readability, though, we will follow common practice and usually drop outermost pairs of brackets.

From these symbols, we can construct the *terms* of L_B. A term is a referring expression built up from occurrences of '0' and applications of the function expressions 'S', '+', '×'. Examples are 0, SSS0, S0 + SS0, (S0 + SS0) × SSS0, SSS0 + ((S0 + SS0) × SSS0), and so on.

We will use σ and τ throughout this chapter as metalinguistic placeholders for terms. The *value* of a term τ is the number it denotes when standardly interpreted: the values of our example terms are respectively 0, 3, 3, 9 and 12.

(b)   The sole built-in predicate of the language L_B is the identity sign. Since L_B lacks non-logical predicates, the only way of forming atomic wffs in the language is therefore by taking two terms and putting the identity sign between them. In other words, the atomic wffs of L_B are *equations* relating terms denoting particular numbers. So, for example, S0 + SS0 = SSS0 is a true atomic wff – which we can abbreviate as $\overline{1} + \overline{2} = \overline{3}$. And S0 + SS0 = SS0 × SS0 is a false atomic wff – which we can abbreviate as $\overline{1} + \overline{2} = \overline{2} \times \overline{2}$.

We now add a negation sign to the language L_B so that we can also explicitly assert that various equations do *not* hold. For example, $\neg$ S0 + SS0 = SS0 × SS0 is true. Though, for readability's sake, we will prefer to rewrite a wff of the form $\neg\sigma = \tau$ as $\sigma \neq \tau$, so that last wff becomes S0 + SS0 $\neq$ SS0 × SS0.

We will also give L_B the conditional connective.

6.2   The axioms and logic of Baby Arithmetic

(a)   The theory BA couched in this language L_B will come equipped with a classical deductive system to deal with negation, the conditional and identity.

We can take the principles governing the connectives to be familiar. And to deal with identity, we need the principle that any sentence of the form $\tau = \tau$ is a logical truth, together with *Leibniz's Law* which allows us to intersubstitute identicals – in other words, given $\varphi(\tau)$ and either $\sigma = \tau$ or $\tau = \sigma$, we can infer $\varphi(\sigma)$. In a handful of illustrations, we'll set out proofs in a Fitch-like natural deduction format (because it is likely to be familiar, and is in any case easy to follow): nothing hangs on this choice of logical system.

(b)   Next, we want non-logical axioms governing the successor function. We want to capture the idea that, if we start from zero and repeatedly apply the successor function, we keep on getting further numbers – i.e. different numbers have different successors: contraposing, for any m, n, if $Sm = Sn$ then $m = n$. Further, zero isn't a successor, i.e. we never cycle back to zero: for any n, $0 \neq Sn$.

However, there are no quantifiers in L_B. So we can't directly express those general facts about the successor function inside the object language L_B. Rather, we have to employ *schemas* (i.e. general templates) and use the generalizing apparatus in our English metalanguage. So we say *any sentence that you get from one of the following schemas by substituting standard numerals for the place-holders 'ζ', 'ξ' is an axiom:*

Schema 1 $0 \neq S\zeta$

Schema 2 $S\zeta = S\xi \rightarrow \zeta = \xi$

NB: These schemas are *not* themselves axioms of BA; the Greek metavariables don't belong to the language L_B. It is, to repeat, *instances* of the schemas got by systematically replacing the placeholders with numerals – same placeholder, same replacement – which are the axioms.[1] We'll see some examples in a moment.

(c)   Similarly, we want non-logical axioms for addition that capture some key ideas underlying the school-room rules. So first we claim that adding zero to a number makes no difference: for any m, $m + 0 = m$. And next, adding a non-zero number Sn (i.e. $n+1$) to m is governed by the following rule: for any m, n, $m + Sn = S(m+n)$ – i.e. $m+(n+1) = (m+n)+1$. These two principles together tell us how to add zero to a given number m; and then adding one is defined as the successor of the result of adding zero; and then adding two is defined as the successor of the result of adding one; and so on up – thus defining adding n for any particular natural number n.

  Because of L_B's lack of quantifiers, we again can't express all that directly inside L_B itself. We have to resort to schemas, and say that anything you get by substituting standard numerals for placeholders in one of the following schemas is an axiom – for short, *every numeral instance of these schemas is an axiom*:

Schema 3 $\zeta + 0 = \zeta$

Schema 4 $\zeta + S\xi = S(\zeta + \xi)$

We can similarly pin down the multiplication function by requiring that *every numeral instance of these schemas too is an axiom*:

Schema 5 $\zeta \times 0 = 0$

Schema 6 $\zeta \times S\xi = (\zeta \times \xi) + \zeta$

Instances of Schema 5 tell us the result of multiplying by zero. Instances of Schema 6 with 'ξ' replaced by '0' define how to multiply by one in terms of first multiplying by zero and then applying the already-defined addition function. Once we know about multiplying by one, we can use another instance of Schema 6 – this time with 'ξ' replaced by 'S0' – to tell us how to multiply by two (multiply by one and then do some addition). And so on, thus defining multiplication for every number.

  To summarize, then,

Defn. 21 BA, *Baby Arithmetic, is the theory whose language is L_B, whose logic comprises classical rules for negation, the conditional, and identity, and whose non-logical axioms are every numeral instance of Schemas 1 to 6.*

So although BA's non-logical axioms fall into just six kinds, there are an infinite number of them – since *any* instance of our schemas counts as an axiom.

---

[1]Fine print: here and below, it wouldn't actually make any difference to the strength of our theory if we allowed the placeholding metavariables to be systematically replaced by any terms, not just by standard numerals. But let's keep things simple.

  Note that *here* we can't use free variables as placeholders – compare §4.2 fn.1. For L_B doesn't have any variables we can recruit for this duty.

However, although it isn't *finitely* axiomatized, it is still an *effectively* axiomatized theory: given a candidate wff, it is clear we can effectively decide whether it is an instance of one of those six schemas and hence an axiom.[2]

(d)   A final remark. Suppose we had adopted versions of our everyday English numerals to denote numbers in a formal arithmetic. Then we would also need a whole bunch of additional axioms like S zero = one, S one = two, and so on, and we would need further school-room rules to handle our base-ten notation. We have avoided this sort of complication by choosing to use our simple-though-long-winded standard numerals as our method of denoting numbers in BA. That's a non-trivial choice but it is this which enables our axioms for BA to be so delightfully simple.

6.3   Proofs of equations inside BA

Let's start with three brisk examples of how arithmetic can be done inside BA, breaking down some informal calculations into minimal steps. These examples are decidedly unexciting: but arguing 'Here are some BA derivations of equations, and we can obviously generalize from these particular cases to get Theorem 10, our next theorem' will actually be more illuminating than giving an abstract general proof.

First, let's show that $\mathsf{BA} \vdash 0 + \overline{2} = \overline{2}$. In other words, $0 + \mathsf{SS0} = \mathsf{SS0}$ is a theorem – and note carefully, this wff *isn't* an instance of Schema 3.

1.	$0 + 0 = 0$	Axiom, instance of Schema 3
2.	$0 + \mathsf{S0} = \mathsf{S}(0 + 0)$	Axiom, instance of Schema 4
3.	$0 + \mathsf{S0} = \mathsf{S0}$	From 1, 2 by Leibniz's Law (LL)
4.	$0 + \mathsf{SS0} = \mathsf{S}(0 + \mathsf{S0})$	Axiom, instance of Schema 4
5.	$0 + \mathsf{SS0} = \mathsf{SS0}$	From 3, 4 by LL

Similarly, we can prove $\overline{2} + \overline{2} = \overline{4}$, i.e. $\mathsf{SS0} + \mathsf{SS0} = \mathsf{SSSS0}$:

1.	$\mathsf{SS0} + 0 = \mathsf{SS0}$	Axiom, instance of Schema 3
2.	$\mathsf{SS0} + \mathsf{S0} = \mathsf{S}(\mathsf{SS0} + 0)$	Axiom, instance of Schema 4
3.	$\mathsf{SS0} + \mathsf{S0} = \mathsf{SSS0}$	From 1, 2 by LL
4.	$\mathsf{SS0} + \mathsf{SS0} = \mathsf{S}(\mathsf{SS0} + \mathsf{S0})$	Axiom, instance of Schema 4
5.	$\mathsf{SS0} + \mathsf{SS0} = \mathsf{SSSS0}$	From 3, 4 by LL

And now let's show that $\mathsf{BA} \vdash \overline{2} \times \overline{2} = \overline{4}$. In unabbreviated form, we need (rather laboriously!) to derive $\mathsf{SS0} \times \mathsf{SS0} = \mathsf{SSSS0}$:

---

[2]More fine print. We definitely want negation in our language of Baby Arithmetic, as we want to be able to formally express true inequalities such as $1 + 1 \neq 3$. But the use of the conditional is in fact optional. As should become clear, we could for our purposes trade in the schema that says that every wff of the form $\mathsf{S}\zeta = \mathsf{S}\xi \rightarrow \zeta = \xi$ is true for a corresponding inference rule that tells us that from a wff of the form $\mathsf{S}\zeta = \mathsf{S}\xi$ we can infer the corresponding wff of the form $\zeta = \xi$. Not having the conditional in play would make the proof of Theorem 14 one step simpler; but it would slightly obscure the point that the key move between Baby Arithmetic and Robinson Arithmetic is adding the quantifiers.

1. $SS0 \times 0 = 0$ Axiom, instance of Schema 5
2. $SS0 \times S0 = (SS0 \times 0) + SS0$ Axiom, instance of Schema 6
3. $SS0 \times S0 = 0 + SS0$ From 1, 2 by LL
4. $0 + SS0 = SS0$ Derived as in first proof above
5. $SS0 \times S0 = SS0$ From 3, 4 by LL
6. $SS0 \times SS0 = (SS0 \times S0) + SS0$ Axiom, instance of Schema 6
7. $SS0 \times SS0 = SS0 + SS0$ From 5, 6 by LL
8. $SS0 + SS0 = SSSS0$ Derived as in second proof above
9. $SS0 \times SS0 = SSSS0$ From 7, 8 by LL

OK: so now let's generalize. Suppose that for some other m we'd started instead from the Axiom $\overline{m} + 0 = \overline{m}$, another instance of Schema 3. Then by similar steps as for the first two proofs, we can derive $\overline{m} + SS0 = SS\overline{m}$, i.e. $\overline{m} + \overline{2} = \overline{m+2}$ (here, $\overline{m+2}$ of course stands in for the standard numeral for $m + 2$).

And then, generalizing further, if we keep extending the same proof idea with a few more steps cut to the same pattern, we can get BA to show $\overline{m} + \overline{3} = \overline{m+3}$, and $\overline{m} + \overline{4} = \overline{m+4}$, and so on. In fact, for any m, n, $\text{BA} \vdash \overline{m} + \overline{n} = \overline{m+n}$.

Next, looking at our third sample proof, we see that we'll be able to similarly prove $\overline{m} \times \overline{2} = \overline{m \times 2}$ for any m. And then, generalizing further, if we keep extending the same proof idea with more steps cut to the same pattern, we can prove $\overline{m} \times \overline{3} = \overline{m \times 3}$, and $\overline{m} \times \overline{4} = \overline{m \times 4}$, and so on. In fact, for any m, n, $\text{BA} \vdash \overline{m} \times \overline{n} = \overline{m \times n}$.

We can now generalize a step further: BA can correctly evaluate not just the simplest terms but *all* terms of its language. That is to say,

Theorem 10 *Suppose τ is a term of L_B and suppose the value of τ on the intended interpretation of the symbols is t. Then $\text{BA} \vdash \tau = \bar{t}$.*

Proof Let's take a very simple example and then draw the required general conclusion. So suppose we want to show e.g. that $(\overline{2} + \overline{3}) \times (\overline{2} \times \overline{2}) = \overline{20}$ – you'll forgive me for not writing out '$\overline{20}$' in basic notation with its twenty occurrences of 'S'! Then we can proceed as follows, again arguing in BA:

1. $(\overline{2} + \overline{3}) \times (\overline{2} \times \overline{2}) = (\overline{2} + \overline{3}) \times (\overline{2} \times \overline{2})$ Identity law
2. $\overline{2} + \overline{3} = \overline{5}$ BA can do simple addition
3. $(\overline{2} + \overline{3}) \times (\overline{2} \times \overline{2}) = \overline{5} \times (\overline{2} \times \overline{2})$ From 1, 2 by LL
4. $\overline{2} \times \overline{2} = \overline{4}$ BA can do simple multiplication
5. $(\overline{2} + \overline{3}) \times (\overline{2} \times \overline{2}) = \overline{5} \times \overline{4}$ From 3, 4 by LL
6. $\overline{5} \times \overline{4} = \overline{20}$ BA can do simple multiplication
7. $(\overline{2} + \overline{3}) \times (\overline{2} \times \overline{2}) = \overline{20}$ From 5, 6 using LL

What we do here is 'evaluate' the complex formula on the right 'from the inside out', reducing the complexity of what's on the right at each stage, and hence eventually equating the complex formula on the left with a standard numeral on the right. Evidently, we can always do this trick, whatever complex formula we start from. So the desired general claim follows. ⊠

From this last result, we can immediately deduce

Theorem 11 *If $\sigma = \tau$ is a true equation, then* $\mathsf{BA} \vdash \sigma = \tau$.

Proof If $\sigma = \tau$ is true, then σ and τ must evaluate to the same number n. Hence by Theorem 10, we have both $\mathsf{BA} \vdash \sigma = \bar{n}$ and $\mathsf{BA} \vdash \tau = \bar{n}$. From which it immediately follows that $\mathsf{BA} \vdash \sigma = \tau$ by Leibniz's Law. ⊠

6.4 Proofs of inequations inside BA

Next, we note that BA knows that different standard numerals are indeed not equal. For example, here's a BA proof of $\bar{4} \neq \bar{2}$:

1.	SSSS0 = SS0	Supposition
2.	SSSS0 = SS0 → SSS0 = S0	Axiom, instance of Schema 2
3.	SSS0 = S0	From 1, 2 by Modus Ponens
4.	SSS0 = S0 → SS0 = 0	Axiom, instance of Schema 2
5.	SS0 = 0	From 3, 4 by Modus Ponens
6.	0 ≠ SS0	Axiom, instance of Schema 1
7.	Contradiction!	From 5, 6 and identity rules
8.	SSSS0 ≠ SS0	From 1 to 7, by Reductio ad Absurdum.

And a little reflection on this illustrative proof should now convince you that in general

Theorem 12 *If s and t are distinct numbers, then* $\mathsf{BA} \vdash \bar{s} \neq \bar{t}$.

And that immediately gives us a companion result to Theorem 11:

Theorem 13 *If $\sigma = \tau$ is a false equation, then* $\mathsf{BA} \vdash \sigma \neq \tau$.

Proof If $\sigma = \tau$ is false, then σ will evaluate to s and τ will evaluate to t where $s \neq t$. So by Theorem 11, we have both $\mathsf{BA} \vdash \sigma = \bar{s}$ and $\mathsf{BA} \vdash \tau = \bar{t}$, and by Theorem 12 $\mathsf{BA} \vdash \bar{s} \neq \bar{t}$. So two applications of Leibniz's Law give us the desired result, $\mathsf{BA} \vdash \sigma \neq \tau$. ⊠

6.5 BA is a sound and negation-complete theory of the truths of L_B

Theorems 11 and 13 tell us that, as far as the atomic wffs of L_B are concerned (i.e. equations of the form $\sigma = \tau$), the true ones are provable in BA and the false ones are refutable. So we could say that BA is negation complete for the atomic wffs of L_B. And we can now go on to show that the same holds for *all* wffs of L_B. We just need to appeal to a basic fact about propositional logic to derive

Theorem 14 BA *is a sound effectively axiomatized theory which is negation complete.*

Proof BA is obviously a sound theory – its axioms are trivial arithmetical truths, and its logic is truth-preserving. It is equally obviously an effectively axiomatized theory, so all its theorems are true. So we just need to tackle the issue of negation completeness.

33

Start with a general observation. A wff φ of L_B is built using negation and other truth-functional connectives from some atomic wffs (in fact, equations) $\alpha_1, \alpha_2, \ldots, \alpha_n$. So now consider any assignment V of truth-values to those atoms. Let α_i^V simply be α_i if α_i is true on V, and be $\neg\alpha_i$ otherwise. Similarly, let φ^V simply be φ if φ is true on V, and be $\neg\varphi$ otherwise. Then, just by the definition of the α_i^V and φ^V, the one and only way of making the α_i^V all true together (i.e. valuation V) makes φ^V true. Hence $\alpha_1^V, \alpha_2^V, \ldots, \alpha_n^V$ together tautologically entail φ^V. Hence by the completeness theorem for a standard propositional logic PL, we have (∗) $\alpha_1^V, \alpha_2^V, \ldots, \alpha_n^V \vdash_{PL} \varphi^V$.

So now fix on the particular valuation V which gives these atoms (equations) their arithmetically correct values. Then, as we in effect noted before, Theorems 11 and 13 between them tell us that BA $\vdash \alpha_i^V$ for each equation α_i.

But BA contains a propositional logic PL. So, by our result (∗), BA proves φ^V – i.e. it proves whichever of φ and $\neg\varphi$ is the true one. Therefore BA is negation complete. ◻

"Hold on! I thought we couldn't have a sound effectively axiomatized theory of arithmetic which is negation complete." No. Theorem 1 didn't say *that*: it said we couldn't have a sound, negation-complete, effectively axiomatized theory which contains what we called the language of basic arithmetic – and *that* language allows us to quantify over numbers. By contrast, L_B is quantifier-free. This language only allows us to express facts about adding and multiplying particular numbers; it can't express numerical generalizations. That's why it can be complete.

6.6   The language L_A

So far that is all very straightforward, but also rather unexciting.[3] The reason that Baby Arithmetic manages to prove every correct claim that it can express – and is therefore negation complete by our Defn. 8 – is that it can't express very much. In particular, as we just stressed, it can't express any generalizations at all. And so the obvious way to beef up BA into something more expressively competent is to restore the familiar apparatus of quantifiers and variables. That's what we'll do next.

First, then, we define the interpreted *first-order language of basic arithmetic* L_A. We will keep the same non-logical vocabulary as in L_B: so there is still just a single non-logical constant denoting zero, plus the three function-symbols, $S, +, \times$, still expressing successor, addition and multiplication. But now we allow ourselves the full expressive resources of first-order logic; so we now have the

---

[3]Mathematically unexciting, anyway. But there is perhaps some philosophical interest. For we might reasonably suppose that the axiom schemas of BA at least partially encapsulate the meanings of the symbols for zero and for the successor, addition and multiplication functions – they partially define what we are talking about. So it is really rather tempting to be a logicist at least about the arithmetic truths proved by BA, regarding them as truths of logic-plus-definitions. And this success might encourage us to pursue some more ambitious form of logicism (see §2.6).

propositional connectives plus the usual supply of quantifiers and variables to express generality, as well as the built-in identity predicate. We fix the domain of the quantifiers to be the natural numbers. The result is the language L_A: it is the least ambitious language which 'contains the language of basic arithmetic' in the sense of Defn. 11.

6.7 Robinson Arithmetic, Q

With this richer formal language available, we can define *Robinson Arithmetic*, commonly denoted simply 'Q'.[4] This is a theory built in the language L_A. It is equipped with a full proof system for first-order classical logic. And for its non-logical axioms, now that we have the quantifiers available to express generality, we can replace each of BA's metalinguistic schemas (specifying an infinite number of formal axioms governing particular numbers) by a single generalized Axiom expressed inside L_A itself.

For example, we can replace the first two schemas governing the successor function by the following:

Axiom 1 $\forall x(0 \neq Sx)$

Axiom 2 $\forall x \forall y(Sx = Sy \rightarrow x = y)$

Obviously, each instance of our earlier Baby Arithmetic Schemas 1 and 2 can be deduced from the corresponding Robinson Arithmetic Axiom by instantiating the quantifiers with numerals.

These Axioms tell us that zero isn't a successor, but they don't explicitly rule out there being *other* objects that aren't successors cluttering up the domain of quantification. We didn't need to fuss about this before, because by construction BA can only talk about the numbers represented by standard numerals in the sequence '0, S0, SS0, ...'. But now we have the quantifiers in play. And these quantifiers are intended to run over the natural numbers, i.e. over zero and its successors.

So let's add an axiom which says that, other than zero, every number is indeed a successor:

Axiom 3 $\forall x(x \neq 0 \rightarrow \exists y(x = Sy))$

Next, we can similarly replace our previous schemas for addition and multiplication by universally quantified Axioms in the obvious way:

Axiom 4 $\forall x(x + 0 = x)$

Axiom 5 $\forall x \forall y(x + Sy = S(x + y))$

Axiom 6 $\forall x(x \times 0 = 0)$

Axiom 7 $\forall x \forall y(x \times Sy = (x \times y) + x)$

---

[4]The expected 'R' is in fact the name given to a different Robinsonian arithmetic – see fn. 7.

Again, each of these Q axioms entails all the instances of BA's corresponding schema.

In sum, then:

Defn. 22 *The formal theory with language L_A, Axioms 1 to 7, plus a classical first-order logic, is standardly called* Robinson Arithmetic, *or simply* Q.

Since any BA axiom can be derived from one of our new Q Axioms, anything that can be proved in BA can be proved in Q.

6.8   Robinson Arithmetic is not complete

Like BA, Q too is an effectively axiomatized sound theory. Its axioms are all true; and its logic is truth-preserving; so its derivations are genuine proofs in the intuitive sense of demonstrations of truth. Every theorem of Q is a true L_A wff, then. But just which truths of L_A are theorems of Q?

On the positive side,

Theorem 15 Q *correctly decides every quantifier-free L_A sentence. In other words, $Q \vdash \varphi$ if the quantifier-free wff φ is true, and $Q \vdash \neg\varphi$ if the quantifier-free wff φ is false.*

Proof We know that Q (like BA) will correctly decide every atomic wff, i.e. correctly decide every equation between terms. And as in our proof of Theorem 14, it follows that Q must then correctly decide every wff built up from those atoms using just the truth-functional propositional connectives.   ⊠

So far, so good. However, there are very simple true *quantified* sentences that Q can't prove.

For example, while Q can prove any particular wff of the form $0 + \bar{n} = \bar{n}$, *it can't prove the corresponding universal generalization:*

Theorem 16 $Q \nvdash \forall x(0 + x = x)$.

Proof   Since Q is a theory with a standard first-order logic, for any L_A-sentence φ, $Q \vdash \varphi$ only if $Q \vDash \varphi$ (by the soundness theorem for first-order logic).

Put $U =_{\text{def}} \forall x(0 + x = x)$. Then one way of showing that $Q \nvdash U$ is to show that $Q \nvDash U$: and we can show *that* by producing a countermodel to the entailment – i.e. by finding an interpretation (a deviant, unintended, 'non-standard', re-interpretation) for L_A's wffs which makes Q's axioms true-on-that-interpretation but which makes U false.

So here goes: take the domain of our deviant, unintended, re-interpretation to be the set N^* which comprises the natural numbers but with two other 'rogue' elements a and b added (these could be e.g. Kurt Gödel and his friend Albert Einstein – but any other pair of distinct non-numbers will do). Let '0' still refer to zero. And take 'S' now to pick out the successor* function S^* which is defined as follows: $S^*n = Sn$ for any natural number in the domain, while for our rogue elements $S^*a = a$, and $S^*b = b$. It is very easy to check that

Axioms 1 to 3 are still true on this deviant interpretation. Zero is still not a successor. Different elements have different successors. And every non-zero element is again a successor (perhaps a self-successor! – though not necessarily an eventual successor of zero).

We now need to extend this interpretation to re-interpret the function-symbol '+'. Suppose we take this to pick out addition*, where $m +^* n = m + n$ for any natural numbers m, n in the domain, while $a +^* n = a$ and $b +^* n = b$. Further, for any x (whether number or rogue element), $x +^* a = b$ and $x +^* b = a$. If you prefer that in a table, then read off *row $+^*$ column* here:

$+^*$	n	a	b
m	$m + n$	b	a
a	a	b	a
b	b	b	a

It is again easily checked that interpreting '+' in Q as addition* still makes Axioms 4 and 5 true.[5] And note that we have e.g. $0 +^* a \neq a$, so on this interpretation U, i.e. $\forall x(0 + x = x)$, fails.

We are not quite done, however, as we still need to show that we can give a co-ordinate re-interpretation of '×' in Q by some deviant multiplication* function. We can leave it as an exercise to fill in suitable details.

Then, with the details filled in, we will have an overall interpretation which makes the axioms of Q true and U false. So $Q \nvdash U$ ⊠

Theorem 17 Q *is negation incomplete.*

Proof We've just shown that Q can't prove U. But obviously, Q can't prove ¬U either. Revert to the standard interpretation built into L_A. All Q's theorems are true but ¬U is false on that interpretation. So ¬U can't be a theorem. Hence U is formally undecidable in Q. ⊠

Of course, we've already announced that Gödel's incompleteness theorem is going to prove that *no* sound axiomatized theory whose language is at least as rich as L_A can be negation complete – that was Theorem 1. But we don't need to invoke anything remotely as elaborate as Gödel's arguments to see that Q is negation incomplete. Q is, so to speak, *boringly* incomplete.

6.9 Statements of order in Robinson Arithmetic

We found a deviant interpretation of Q's axioms by exploiting the fact that, while Axiom 3 ensures every object other than zero is a successor of something,

---

[5]In headline terms: For Axiom 4, we note that adding* zero on the right always has no effect. For Axiom 5, just consider cases. (i) $m +^* S^*n = m + Sn = S(m + n) = S^*(m +^* n)$ for 'ordinary' numbers m, n in the domain. (ii) $a +^* S^*n = a = S^*a = S^*(a +^* n)$, for 'ordinary' n. Likewise, (iii) $b +^* S^*n = S^*(b +^* n)$. (iv) $x +^* S^*a = x +^* a = b = S^*b = S^*(x +^* a)$, for any x in the domain. (v) Similarly, $x +^* S^*b = S^*(x +^* b)$. Which covers every possibility.

that axiom still allows 'stray objects' which aren't eventual successors of zero. So obviously we'll want to add more axioms to Q in the hope of pinning down the structure of the natural numbers and eliminating 'strays'. That's our task in the next chapter. But let's stick with Robinson Arithmetic for a bit longer. For as we will note in the next section, weak though it is, Q has a property which makes it of considerable interest.

And as a warm-up in this section, we will prove a result that will later be useful.

Theorem 18 *In Robinson Arithmetic, the* less-than-or-equal-to *relation is not just expressed but captured by the wff* $\exists v(v + x = y)$.

Proof It is obvious that the wff expresses the relation. So – recalling the definition of capturing in §5.3, Defn. 19 – what we need to show is that, for any particular pair of numbers, m, n, if $m \leq n$, then $Q \vdash \exists v(v + \overline{m} = \overline{n})$, and if $m > n$, then $Q \vdash \neg \exists v(v + \overline{m} = \overline{n})$.

Suppose $m \leq n$, so for some $k \geq 0$, $k + m = n$. Q can prove everything BA proves and hence, in particular, can prove every true addition equation. So we have $Q \vdash \overline{k} + \overline{m} = \overline{n}$. But then $\exists v(v + \overline{m} = \overline{n})$ follows by existential quantifier introduction. Therefore $Q \vdash \exists v(v + \overline{m} = \overline{n})$, as was to be shown.

Suppose alternatively $m > n$. We need to show $Q \vdash \neg \exists v(v + \overline{m} = \overline{n})$. We'll first demonstrate this in the case where $m = 2$, $n = 1$, using a Fitch-style proof system. For brevity we will omit statements of Q's axioms and some other trivial steps; we drop unnecessary brackets too.

1.	$\exists v(v + SS0 = S0)$	Supposition
2.	$a + SS0 = S0$	Supposition
3.	$a + SS0 = S(a + S0)$	From Axiom 5
4.	$S(a + S0) = S0$	From 2, 3 by LL
5.	$a + S0 = S(a + 0)$	From Axiom 5
6.	$SS(a + 0) = S0$	From 4, 5 by LL
7.	$a + 0 = a$	From Axiom 4
8.	$SSa = S0$	From 6, 7 by LL
9.	$SSa = S0 \rightarrow Sa = 0$	From Axiom 2
10.	$Sa = 0$	From 8, 9 by Modus Ponens
11.	$0 = Sa$	From 10
12.	$0 \neq Sa$	From Axiom 1
13.	Contradiction!	From 11, 12
14.	Contradiction!	$\exists E$ 1, 2–13
15.	$\neg \exists v(v + SS0 = S0)$	From 1–14 by Reductio

The only step to explain may be at line (14) where we use a version of the Existential Elimination rule: if the temporary supposition $\varphi(a)$ leads to contradiction, for arbitrary a, then $\exists v \varphi(v)$ must lead to contradiction.

And having constructed a proof for the case $m = 2$, $n = 1$, inspection reveals that we can use the same general pattern of argument to show $Q \vdash \neg \exists v(v + \overline{m} = \overline{n})$ whenever $m > n$. So we are done. $\boxtimes$

Given the theorem we have just proved, we can sensibly add the standard symbol '$\leq$' to L_A, the language of Q, defined so that for any L_A terms – including of course variables – $\sigma \leq \tau$ is just short for $\exists v(v + \sigma = \tau)$.[6] And then Q will be able to prove at least the expected facts about the less-than-or-equals relations among quantifier-free terms.

Note, by the way, that some presentations treat '$\leq$' as a primitive symbol built into our formal theories like Q from the start, governed by its own additional axiom(s). But nothing important hangs on the difference between that approach and our policy of introducing the symbol by definition.

And of course, nothing hangs either on our policy of introducing '$\leq$' as our basic order symbol rather than '$<$', which could have been defined by $\sigma < \tau =_{\text{def}} \exists v(Sv + \sigma = \tau)$.

6.10   Why Robinson Arithmetic is interesting

(a)   Given it can't even prove $\forall x(0 + x = x)$, Q is evidently a *very* weak theory of arithmetic. Even so, Q does have some very interesting features.

As we just saw, it can capture the particular decidable relation that obtains when one number is at least as big as another. And in fact, we can now announce a surprisingly sweeping general result:

Theorem 19 Q *can capture* all *effectively decidable numerical properties (and relations) – hence it is sufficiently strong in the sense of Defn 20.*

That might initially seem *very* unexpected, given Q's weakness. But remember, 'sufficient strength' was defined as a matter of being able to *case-by-case* prove enough wffs about decidable properties of individual numbers. It turns out that Q's hopeless weakness at proving *generalizations* doesn't stop it from proving enough facts about *particular* numbers.

That's why Q is especially interesting – it is one of the very weakest arithmetics which is sufficiently strong, and it was isolated by Raphael Robinson in 1952 for just that reason.[7] Hence Q is one of the weakest arithmetics for which Gödelian proofs of incompleteness can be run. Suppose, then, that a theory is formally axiomatized, consistent and can prove everything Q can prove (surely very weak requirements). Then what we've just announced and promised can be proved is that any such theory will be sufficiently strong. And therefore e.g. Theorem 9 will apply – any such theory will be incomplete.

(b)   However, we can only prove the announced Theorem 19 that Q *does* have sufficient strength if and when we have a general theory of effective decidability

---

[6]Fine print: strictly, we need to choose the quantified variable to avoid any clash of variables with the substituted terms. We won't fuss about that.

[7]We can't say that Q is *the* weakest sufficiently strong arithmetic. Robinson also isolated another very weak but sufficiently strong arithmetic R which neither contains nor is contained in Q. But Robinson's other theory isn't finitely axiomatized, so it is usual to focus on the prettier and finitely axiomatized Q.

to hand. And as we said at the end of the previous chapter, we don't want to get embroiled yet in developing that theory. So what we *will* be proving quite soon (in Chapter 10) is a somewhat weaker claim about Q. We'll show that it can capture all so-called 'primitive recursive' properties and relations, where these form a large and very important subclass of the effectively decidable ones. This major theorem will be a crucial load-bearing part of our proofs of various Gödel style incompleteness theorems: it means that Q gives us 'the modest amount of arithmetic' needed for a version of Theorem 2.

But before we get round to showing all that, we are first going to take a look at a *much* richer arithmetic than Q, namely PA.

7 First-order Peano Arithmetic

The previous chapter introduced two weak theories of arithmetic, BA and Q. In this chapter – jumping over a whole family of intermediate-strength theories – we introduce a *much* richer first-order theory of arithmetic, PA. It's what you get by adding a generous supply of induction axioms to Q.

7.1 Mathematical induction: the very idea

Here is the basic idea we need (and throughout, take 'number' to mean 'natural number'):

> Whatever numerical property we take, if (i) zero has this property, and also (ii) the property is passed down from any number n which has it to its successor Sn, then it follows that (iii) the property is had by *all* numbers.

This is the key *principle of mathematical induction*, used in very many informal proofs of arithmetical generalizations.[1]

The principle is plainly a sound one, guaranteed by the structure of the number series. Suppose (i) 0 has property P and also (ii) for any n, if n has P so does Sn. Since 0 has P, we can use (ii) to deduce $S0$ has P. Since $S0$ has P, we can use (ii) to deduce that $SS0$ has P. Then, by appeal to (ii) again, it follows that $SSS0$ has P. And so on. Hence (iii) property P will in this way percolate down to any given number – since you can get to any natural number by starting from zero and repeatedly applying the successor function (there are no stray natural numbers, lying outside that sequence).

7.2 The induction axiom, the induction schema

The intuitive idea, to repeat, is that for any property of numbers, if zero has it and if it is passed from one number to the next, then all numbers have it. Putting it this way involves generalizing over properties of numbers. So to frame a corresponding formal induction principle, we might naturally want to use a formalized

---

[1] I'll assume this is very familiar. If not, you can start by looking at the Wikipedia article on 'Mathematical induction', or (better!) look at the chapter with the same title in Daniel J. Velleman's *How to Prove It* (CUP).

language that likewise allows us to generalize over properties of numbers. And the obvious option is to adopt a language with *second-order* quantifiers.

In other words, we will want a language that not only has first-order quantifiers of the familiar sort, in this case running over the natural numbers, but also has a second type of quantifier which runs over all properties-of-numbers. In such a language, we can then state a *second-order Induction Axiom* as follows:

$$\forall X(\{X0 \land \forall x(Xx \rightarrow XSx)\} \rightarrow \forall xXx)$$

You can read this as saying 'For any property X, given both that 0 has X and also that if any number has X so does its successor, then *every* number has property X.'[2]

Natural though this second-order axiom might be, we will however be focusing on formal theories whose logical apparatus involves only first-order quantification over numbers. And we will continue to be particularly interested in theories which are framed in the language L_A.

Why? This isn't due to some perverse desire to work with one hand tied behind our backs. It's because there are troublesome issues about second-order logic. For a start, there are technical issues: second-order logic's consequence relation (at least on the natural understanding) can't be captured in a nice formalizable logical system: hence theories using a full second-order logic aren't effectively axiomatizable. And then there are more philosophical issues: just how well do we really understand the intuitive idea of quantifying over 'all properties of numbers'? Is that really a determinate totality which we can quantify over? We don't want to tangle with such worries here and now.

But if we don't have second-order quantifiers available to range over properties of numbers, how can we handle arithmetical induction in a first-order setting?

Recall how we dealt with a similar problem in the context of Baby Arithmetic. The language L_B lacks first-order quantifiers; so we couldn't use just a single quantified wff to formalize the claim that, for every number n, $n + 0 = n$. Instead, we had to adopt a whole package of unquantified axioms, with each numeral instance of the schema $\zeta + 0 = \zeta$ counting as a separate axiom. We'll now do something analogous here.

The language L_A lacks second-order quantifiers; so we can't write down a single second-order Axiom which directly formalizes the intuitive induction principle. Instead, we will have to adopt a whole package of axioms (which involve only first-order quantifiers) by saying that for each suitable wff $\varphi(x)$, the corresponding instance of the *first-order Induction Schema*

$$\{\varphi(0) \land \forall x(\varphi(x) \rightarrow \varphi(Sx))\} \rightarrow \forall x\varphi(x)$$

is to count as an axiom.[3] Or at least, that is the basic idea. The next section elaborates.

---

[2] Conventionally, we use upper case letters for predicates; correspondingly, we use upper case letters for variables which can occur in place of predicates.

[3] Equivalently, we could adopt a rule of inference to the effect that, given $\varphi(0)$ and $\forall x(\varphi(x) \rightarrow \varphi(Sx))$, we can deduce $\forall x\varphi(x)$.

7.3   Being generous with induction

Suppose then that we are starting from the first-order arithmetic Q, and aim to build a richer theory in its language L_A by adding instances of the Induction Schema as additional axioms. But what makes for a 'suitable' predicate-expression $\varphi(\mathsf{x})$ for use in an instance of the Schema?

Consider *any* open wff $\varphi(\mathsf{x})$ of L_A with one free variable. This will be built up from no more than the constant term '0', the familiar successor, addition and multiplication functions, plus the identity relation and other logical apparatus. Therefore – you might very well suppose – it ought to be a perfectly determinate matter, for each n, whether $\varphi(\bar{\mathsf{n}})$ is true or not. In other words, $\varphi(\mathsf{x})$ ought to express a perfectly determinate arithmetical property of numbers (even if, in the general case, we can't always effectively decide whether a given number n has the property or not). But if $\varphi(\mathsf{x})$ does express a perfectly determinate arithmetical property P, then – by the intuitive argument of §7.1 – we should be able to argue by induction that if zero has P, and if P always gets passed down from a number to its successor, then all numbers have property P. *So why not be generous and allow any open one-variable L_A-wff $\varphi(\mathsf{x})$ to count as suitable for substitution in the Induction Schema?*

But there's more: we will also want to make use of instances of the Induction Schema where $\varphi(\mathsf{x})$ has further variables dangling free.

Why? Consider informal mathematical argumentation again. We often want to establish universally generalized *relational* claims about numbers, and might then need to use an argument of the following overall shape:

> Pick an arbitrary number a. Then (i) 0 is R to a. And (ii) for any m, if m is R to a, then Sm is also R to a. Hence by induction, (iii) for any m, m is R to a. But a was arbitrary. So we can conclude that m is R to n for *any* numbers m, n.

The core part of the argument here is an induction for, so to speak, the property of *being-R-to-a* (where 'a' serves as a parameter, a temporary name for an arbitrarily selected number). So now the question is: how could we frame a formal analogue for the inductive argument here?

The obvious move is to allow instances of the induction schema for a wff $\varphi(\mathsf{x},\mathsf{y})$ with the free variable 'y' serving as a parameter.[4]

Then, if you don't mind the axioms of a theory having free variables, we can simply say as before that each instance of the induction schema – now including those instances with variables dangling free in φ – is to be an axiom. If you do mind, then we can instead say that it is the *closure* of each instance of the induction schema which is an axiom (where, of course, the closure of a wff is what you get by universally quantifying any dangling free variables). I somewhat prefer this second line, because we can then still think of a sound arithmetical theory as always having true *sentences* as axioms.

---

[4]Fine print: If your formal logic insists on typographically distinguishing free-variables-used-parametrically from bound variables, you'd instead use the likes of $\varphi(\mathsf{x},\mathsf{a})$.

7.4 First-order Peano Arithmetic introduced

Suppose then that we are generous with induction and agree that *any* open wff of L_A is suitable for use in an instance of the induction schema. This means moving on from Q, and jumping over a range of possible intermediate theories, to adopt the much richer theory of arithmetic that we can briskly define as follows:

Defn. 23 PA – First-order Peano Arithmetic[5] – *is the theory with a standard first-order logic whose language is L_A and whose axioms are those of Q plus (the closures of) all instances of the Induction Schema that can be constructed from open wffs of L_A.*

PA as presented here has an infinite number of axioms. However that's fine: it is plainly still decidable whether any given wff has the right shape to be one of the new induction axioms, so this is still an effectively formalized theory.

To help fix ideas, then, let's have three easy examples of what we can formally prove using induction.

(a)   First, we'll check that we have plugged the particular gap we noted in Q. Recall: while Q can prove each separate instance of $0 + \bar{n} = \bar{n}$, it very feebly can't prove $\forall x(0 + x = x)$. However, PA can.

How? We just put $0 + x = x$ for $\varphi(x)$, prove $\varphi(0)$, prove $\forall x(\varphi(x) \to \varphi(Sx))$, and use induction to conclude $\forall x \varphi(x)$. Spelling that out in tediously plodding detail (since you insist!):

1.	$0 + 0 = 0$	Instance of Q's Axiom 4
2.	$0 + a = a$	Supposition
3.	$S(0 + a) = Sa$	From 2 by the identity laws
4.	$0 + Sa = S(0 + a)$	Instance of Q's Axiom 5
5.	$0 + Sa = Sa$	From 3, 4
6.	$0 + a = a \to 0 + Sa = Sa$	From 2–5 by Conditional Proof
7.	$\forall x(0 + x = x \to 0 + Sx = Sx)$	From 6, since a was arbitrary.
8.	$\{0 + 0 = 0 \land \forall x(0 + x = x \to 0 + Sx = Sx)\} \to \forall x(0 + x = x)$	
		Instance of Induction Schema
9.	$\forall x(0 + x = x)$	From 1, 7, 8 by propositional logic

(b)   Next, take $\varphi(x)$ to be $x \neq Sx$. Then PA trivially proves $\varphi(0)$ because that's Q's Axiom 1. PA also proves $\forall x(\varphi(x) \to \varphi(Sx))$ by contraposing Axiom 2. And then an induction axiom tells us that if we have both $\varphi(0)$ and $\forall x(\varphi(x) \to \varphi(Sx))$ we can deduce $\forall x \varphi(x)$, i.e. $\forall x\, x \neq Sx$, i.e. no number is a self-successor.

Simple! Yet this trivial little result is worth noting when we recall our deviant interpretation which made the axioms of Q true while making $\forall x(0 + x = x)$ false: that interpretation featured Kurt Gödel himself added to the domain as a rogue self-successor. A bit of induction, however, rules out self-successors.

---

[5] The name is conventional. Giuseppe Peano did publish a list of axioms for arithmetic in 1889. But they weren't first-order, only explicitly governed the successor relation, and – as Peano acknowledged – had already been stated by Richard Dedekind.

(c) A third observation. PA allows, in particular, induction for the formula $\varphi(\mathsf{x}) =_{\text{def}} (\mathsf{x} \neq 0 \rightarrow \exists \mathsf{y}(\mathsf{x} = \mathsf{Sy}))$.

But now note that the corresponding $\varphi(0)$ is a trivial theorem. $\forall \mathsf{x}\varphi(\mathsf{Sx})$ is an equally trivial theorem (why?), and that logically entails $\forall \mathsf{x}(\varphi(\mathsf{x}) \rightarrow \varphi(\mathsf{Sx}))$. So we can use an instance of the Induction Schema inside PA to derive $\forall \mathsf{x}\varphi(\mathsf{x})$.

However, that's just Axiom 3 of Q. Our initial presentation of PA – as explicitly having all the Axioms of Q – therefore involves a certain redundancy.

7.5 Summary overview of PA

Given its very natural motivation, PA is the benchmark axiomatized first-order theory of basic arithmetic. It is probably worth pausing, then, to bring together all the elements of its specification in one place.

First, to repeat, the *language* of PA is L_A, a first-order language whose non-logical vocabulary comprises just the constant '0', the one-place function symbol 'S', and the two-place function symbols '+', '×'. The built-in interpretation for L_A gives those symbols their familiar interpretation in elementary arithmetic and construes the quantifiers as running over the natural numbers.

Second, PA's deductive *proof system* is some standard version of classical first-order logic with identity. The differences between various presentations of the logic of course don't make a difference to what sentences can be proved in PA. We just sketched a proof using a Fitch-style system. Though, as we will see, it will soon be convenient to fix officially on a linear Hilbert-style axiomatic system for later metalogical work theorizing about PA.

And third, PA's *non-logical axioms* – eliminating the redundancy we just noted from our original specification – are the following sentences:

Axiom 1 $\forall \mathsf{x}(0 \neq \mathsf{Sx})$

Axiom 2 $\forall \mathsf{x}\forall \mathsf{y}(\mathsf{Sx} = \mathsf{Sy} \rightarrow \mathsf{x} = \mathsf{y})$

Axiom 3 $\forall \mathsf{x}(\mathsf{x} + 0 = \mathsf{x})$

Axiom 4 $\forall \mathsf{x}\forall \mathsf{y}(\mathsf{x} + \mathsf{Sy} = \mathsf{S}(\mathsf{x} + \mathsf{y}))$

Axiom 5 $\forall \mathsf{x}(\mathsf{x} \times 0 = 0)$

Axiom 6 $\forall \mathsf{x}\forall \mathsf{y}(\mathsf{x} \times \mathsf{Sy} = (\mathsf{x} \times \mathsf{y}) + \mathsf{x})$

plus (the closure of) every instance of the

Induction Schema $\{\varphi(0) \wedge \forall \mathsf{x}(\varphi(\mathsf{x}) \rightarrow \varphi(\mathsf{Sx}))\} \rightarrow \forall \mathsf{x}\varphi(\mathsf{x})$

where $\varphi(\mathsf{x})$ is an open wff of L_A that has 'x' (and perhaps other variables) free.

7.6 What can PA prove?

(a) Even BA is good at proving quantifier-free equations. Q adds some ability to prove quantified wffs. We have so far noted just three additional simple quantified theorems that PA can prove. Further exploration would reveal that PA can prove

45

a whole multitude of other familiar and not-so-familiar basic truths about the successor, addition, multiplication functions – and about the ordering relation (as defined in §6.9). In fact, it can prove so many L_A truths that we might very reasonably have hoped – at least before we'd heard of Gödel's incompleteness theorems – that PA would turn out to be a *negation complete* theory that indeed pins down *all* such truths.

(b) Here is something else that would have encouraged this false hope, pre-Gödel. Suppose we define the language L_P to be L_A without the multiplication sign. Take P – so-called *Presburger Arithmetic* – to be the theory couched in the language L_P, whose axioms are Q's now familiar axioms for successor and addition, plus (the universal closures of) all instances of the induction schema that can be formed in the language L_P. In short, P is PA minus multiplication. *Then* P *is a negation-complete theory of successor and addition.* (We are not going to be able to prove that here – the argument uses a standard model-theoretic method called 'elimination of quantifiers' which isn't hard, and was known in the 1920s, but it would just take a bit too long to explain.)

(c) So the situation is as follows, and was known before Gödel proved his incompleteness theorem. (i) There is a complete formal axiomatized theory BA whose theorems are all the truths about successor, addition and multiplication expressible in the quantifier-free language L_B. (ii) There is another complete formal axiomatized theory P whose theorems are exactly the first-order truths expressible using just successor and addition. Against this background, the result that adding multiplication in order to get full PA gives us a theory which is incomplete and incompletable (if consistent) comes as a rather nasty surprise. It wasn't obviously predictable that adding multiplication was going to make all the difference. Yet it does. Indeed, as we've said before, as soon we have an effectively axiomatized arithmetic as strong as Q which has multiplication as well as addition, we get incompleteness.

And by the way, it isn't that a quantified theory of multiplication must in itself be incomplete. In 1929, assuming we have a complete logic, Thoralf Skolem showed that there is a complete theory for the truths expressible in a suitable first-order language with multiplication but lacking addition or the successor function. So, when quantifiers are in play, why does putting multiplication together with addition and successor produce incompleteness? The answer will emerge shortly enough, but pivots on the fact that even a weak first-order arithmetic like Q with all three functions available can express and capture *all* 'primitive recursive' functions. But we'll have to wait until the next-but-one chapter to explain what *that* means.

7.7   Non-standard models of PA?

(a) I've spoilt any surprise! I've already said that despite its richness, PA is going to turn out to be an incomplete theory. We'll be able to find a Gödel sentence G which is true on the intended interpretation built into the theory's language L_A, but such that PA can't prove it nor indeed disprove it.

PA's logic is complete – which means that the theory can formally prove every sentence which is semantically entailed by the axioms. Hence, since G isn't provable, it can't be semantically entailed by the theory. Which means that there must be some interpretation of the language which still makes all of PA's axioms true but which makes G false. This can't be the standard interpretation built into L_A which makes PA true and makes G true too: hence, alongside its standard model, PA must have non-standard models.[6]

But actually, we don't need to appeal to Gödelian incompleteness to show *that*. For it quickly follows from two elementary theorems of model theory that PA (assuming all along that it is consistent) not only has non-standard models, but even has non-standard countable models – i.e. non-standard models whose domains are no bigger than the set of natural numbers.

(b)   The proof of this last claim is an optional extra. If the two theorems I'm about to state are unfamiliar, no matter – nothing in later chapters depends on following the short argument below. But it is fun, and worth knowing about.

So, the standard bits of model-theory we need are:[7]

Theorem 20 *(Downward) Löwenheim-Skolem. If the theory T (in a first-order language) has an infinite model at all, then it has a countable model.*

Theorem 21 *Compactness. Let Σ be a set of sentences (in a first-order language): if every finite subset of Σ has a model, then so does Σ itself.*

And now, to get our argument going, suppose that we add to the language of PA a new constant c, and add to the axioms of PA the new axioms $c \neq 0$, $c \neq \overline{1}$, $c \neq \overline{2}$, ..., $c \neq \overline{n}$, ..., giving us the augmented theory PA$^+$.

So we have added an infinite set of axioms which, taken all together, in effect tell us that there is a rogue object c (to be denoted by c) which is not an eventual successor of zero. But still, each *finite* subset of the axioms of our new theory PA$^+$ has a model assuming PA does. Just augment the intended standard model of arithmetic by interpreting c as denoting some number greater than the maximum n for which $c \neq \overline{n}$ is in that given finite subset of axioms.

Since each finite subset of the infinite set of axioms of the expanded theory PA$^+$ has a model, the Compactness theorem tells us the *whole* theory must have a model. This model evidently has to be infinite, so by the Löwenheim-Skolem theorem there will be in particular a countable model of this theory. This countable model will contain a 'zero', will have objects to be its eventual 'successors', but also have a rogue object c as the denotation of c (an object which isn't one of the eventual 'successors' of the 'zero'). So this model must

---

[6] Jargon reminder: we say an interpretation for a theory T (or for a set of sentences Σ) is a model of T (for Σ) iff it makes all T's theorems (all the sentences in Σ) true.

[7] These are easy corollaries of the usual completeness proof for first-order logic, and can be found in the standard textbooks.

be structured differently from the standard model of PA. However, since this countable structure is a model of PA-plus-some-extra-axioms it is, a fortiori, a model of PA. Hence PA has a non-standard countable model.[8]

Now, a theory can be negation-complete and yet still have multiple models that look different from each other. For example we mentioned in the last section that Presburger Arithmetic is negation-complete; but it has non-standard models by the same compactness argument. Hence the fact that PA has non-standard models doesn't by itself rule out its being complete. That's why, to establish incompleteness, we will need to get down to our Gödelian arguments!

---

[8]Two points arising. First, this line of argument will apply equally to other theories including ones stronger than PA: i.e. having unintended, non-standard, models is not a special shortcoming of PA.

Second, you might reasonably wonder what one of these countable-but-weird models of PA will look like. In particular, what will be the deviant interpretations of the successor, addition and multiplication functions? Well, the relevant functions do take some effort to describe. For Tennenbaum's Theorem tells us that, for any non-standard model of PA, the interpretations of the addition and the multiplication functions can't be nice computable functions. But pursuing this further would take us too far off-piste.

8 Quantifier complexity

Wffs of the language L_A come in different degrees of *quantifier complexity*. We can distinguish, for a start, so-called Δ_0, Σ_1, and Π_1 wffs. You need to know about these different types of wff for future purposes. This chapter defines them, and presents some initial results.

8.1 Bounded quantifications

We often want to say that all/some numbers less than or equal to some bound have a certain property.

We can express such claims in formal arithmetics like Q and PA by using wffs of the shape $\forall x(x \leq \tau \rightarrow \varphi(x))$ and $\exists x(x \leq \tau \wedge \varphi(x))$. Here the term τ will specify the bound; $\varphi(x)$ expresses the relevant property; and we can treat $x \leq \tau$ as short for $\exists v(v + x = \tau)$ – see §6.9 (we assume τ doesn't contain v free).

It is standard to further abbreviate such wffs by $(\forall x \leq \tau)\varphi(x)$ and $(\exists x \leq \tau)\varphi(x)$ respectively: call the prefixes here *bounded quantifiers*.

And now note that we have easy results like these:

(1) For any n, $Q \vdash \forall x(\{x = \overline{0} \vee x = \overline{1} \vee \ldots \vee x = \overline{n}\} \leftrightarrow x \leq \overline{n})$.

(2) For any n, if $Q \vdash \varphi(\overline{0}) \wedge \varphi(\overline{1}) \wedge \ldots \wedge \varphi(\overline{n})$, then $Q \vdash (\forall x \leq \overline{n})\varphi(x)$.

(3) For any n, if $Q \vdash \varphi(\overline{0}) \vee \varphi(\overline{1}) \vee \ldots \vee \varphi(\overline{n})$, then $Q \vdash (\exists x \leq \overline{n})\varphi(x)$.

Such results show that Q – and hence a stronger theory like PA – 'knows' that bounded universal quantifications with fixed numeral bounds behave like finite conjunctions, and that bounded existential quantifications with fixed numeral bounds behave like finite disjunctions.

8.2 Δ_0 wffs

(a)   A standard definition:

Defn. 24 *An L_A wff is Δ_0 iff it can be built up from the non-logical vocabulary of L_A plus $\leq$ (defined as before), using the familiar propositional connectives and the identity sign, but only using* bounded *quantifiers.*

A Δ_0 wff, then, is just like a quantifier-free L_A wff, except that (i) we are now allowed the existential quantifiers used in unpacking occurrences of $\leq$, and

(ii) we can also allow ourselves to wrap up some finite conjunctions into bounded universal quantifications, and similarly wrap up some finite disjunctions into bounded existential quantifications.

So a Δ_0 wff remains very like a quantifier-free wff; it should therefore come as no surprise to hear this:

Theorem 22 *We can effectively decide the truth-value of any Δ_0 sentence.*

I won't give a full-dress proof. But, roughly speaking, the plan is this. Since we are dealing with *sentences*, outermost bounded quantifiers will be of the type $(\forall x \leq \overline{n})$ or $(\exists x \leq \overline{n})$ so can be unpacked as equivalent to conjunctions or disjunctions. Keep on going, working inwards to eliminate all the bounded quantifiers. And then we are left with an equivalent wff built up using just propositional connectives from basic expressions of the form $\sigma = \tau$ and $\sigma \leq \tau$, where σ and τ are terms containing only numerals (with no variables dangling free). But we can settle the truth values of such basic expressions using simple arithmetic to evaluate the terms, and then we can use truth-tables to determine the values of wffs built up from the basic expressions using the connectives.

Since we can mechanically decide whether a Δ_0 sentence $\varphi(\overline{n})$ is true, this means that we can mechanically determine whether a Δ_0 open wff $\varphi(x)$ is satisfied by a given number n. In other words, a Δ_0 open wff $\varphi(x)$ will express a decidable property of numbers. In the same way, a Δ_0 open wff $\varphi(x, y)$ will express a decidable numerical relation.

(b)   Now, (i) Theorem 15 tells us that even Q can correctly decide all quantifier-free L_A sentences; (ii) Theorem 18 tells us that Q also knows about the relation $\leq$; and (iii), as remarked in the previous section, Q knows that quantifications with fixed number bounds behave just like conjunctions/disjunctions. So the next result won't be a surprise either: what *we* can do informally to decide the truth-value of any Δ_0 sentence, Q can do too!

Theorem 23 Q *(and hence* PA*) can correctly decide all Δ_0 sentences.*

Again, we won't spell out the details of the argument here. But the proof-idea is easy: the simplest Δ_0 sentences are decidable, so we just need to check that building up more complex ones using connectives and bounded quantifiers preserves decidability.

8.3 Σ_1 and Π_1 wffs

(a)   Aside from the quantifiers used in defining '$\leq$', Δ_0 wffs only involve bounded quantifiers. At the next grade up in quantifier complexity, we find the so-called Σ_1 and Π_1 wffs. As our first-shot definitions, we can say that the Σ_1 wffs are what you get by existentially quantifying Δ_0 wffs zero, one, or many times; while the Π_1 wffs are what you get by universally quantifying Δ_0 wffs.[1]

---

[1] The 'Σ' comes from an alternative symbol for the existential quantifier, as in $\Sigma x F x$ – that's a Greek 'S' for '(logical) sum'. Similarly, the 'Π' comes from the corresponding symbol

Why are these notions of Σ_1 and Π_1 wffs of interest? Because – looking ahead – it turns out that Σ_1 wffs are exactly what we need to express and capture computable numerical functions and effectively decidable properties of numbers. And we will find that the standard Gödel sentence that sort-of-says 'I am un-provable' is a Π_1 wff. This means that, while the Gödel sentence might be long and messy, there is also a good sense in which it is logically quite simple.

(b)   Now, the general idea of adding existential (universal) quantifiers to Δ_0 wffs to get Σ_1 (Π_1) wffs is standard. However, official definitions do vary across different textbook presentations. Don't be fazed by this: given enough arithmeti-cal background theory in place, the variants come to the same.

In fact, I'm officially going to plump here for definitions which are rather more involved than our first-shot efforts which had the quantifiers simply added to the front. Sorry! – my excuse is that this will make life a bit easier later on.

We start as follows:

Defn. 25 *We build up the class of strict Σ_1 wffs as follows:*

(1) Any Δ_0 wff is a strict Σ_1 wff.

(2) If φ and ψ are strict Σ_1 wffs, so are $(\varphi \wedge \psi)$ and $(\varphi \vee \psi)$.

(3) If φ is a strict Σ_1 wff, so are $(\forall \xi \leq \kappa)\varphi$ and $(\exists \xi \leq \kappa)\varphi$, where ξ is any variable free in φ, and κ is a numeral or a variable distinct from ξ.

(4) If φ is a strict Σ_1 wff, so is $(\exists \xi)\varphi$ where ξ is any variable free in φ.

(5) Nothing else is a strict Σ_1 wff.

Two comments. First, note the absence of the negation $\neg\varphi$ from clause (2) (as well as the related absence of the conditional $\varphi \to \psi$, i.e. $\neg\varphi \vee \psi$). But of course, if we allowed first applying an (unbounded) existential quantifier to a wff and then negating the result, that would be tantamount to allowing (unbounded) universal quantifications, which isn't at all what we want in defining Σ_1 wffs. Second, there is a certain redundancy in this definition, because it is immediate that clause (4) covers the existential half of clause (3).

The companion definition of the class of strict Π_1 wffs then goes exactly as you would expect – the key change is in clause (4):

Defn. 25 *(continued) And we build up the class of strict Π_1 wffs as follows:*

(1) Any Δ_0 wff is also a strict Π_1 wff.

(2) If φ and ψ are strict Π_1 wffs, so are $(\varphi \wedge \psi)$ and $(\varphi \vee \psi)$.

(3) If φ is a strict Π_1 wff, so are $(\forall \xi \leq \kappa)\varphi$ and $(\exists \xi \leq \kappa)\varphi$, where ξ is any variable free in φ, and κ is a numeral or a variable distinct from ξ.

(4) If φ is a strict Π_1 wff, so is $(\forall \xi)\varphi$ where ξ is any variable free in φ.

(5) Nothing else is a strict Π_1 wff.

---

for the universal quantifier, as in $\Pi x F x$ – that's a Greek 'P' for '(logical) product'. The subscript '1' indicates the first level of a complexity hierarchy which needn't concern us here.

Finally – simply to avoid later having to repeatedly use phrases like 'is strictly Σ_1 or is a logical equivalent' – it will be convenient to add the following:

Defn. 25 *(concluded) A wff is Σ_1 iff it is logically equivalent to a strict Σ_1 wff; and a wff is Π_1 iff it is logically equivalent to a strict Π_1 wff.*

(c)   Let's have a few quick examples, which will also enable us to introduce another bit of terminology. First, put

$$\psi(x) =_{\text{def}} (\exists v \leq x)(2 \times v = x),$$
$$\chi(x) =_{\text{def}} \{x \neq 1 \,\wedge\, (\forall u \leq x)(\forall v \leq x)(u \times v = x \,\rightarrow\, (u = 1 \,\vee\, v = 1))\}.$$

Evidently, $\psi(x)$ expresses the property of being an *even* number. And $\chi(x)$ expresses the property of being *prime* (where we rely on the trivial fact that a number's factors can be no greater than it). Both these wffs are Δ_0.

The sentence

$$\exists x(\psi(x) \wedge \chi(x))$$

says that there is an even prime; it is (strictly) Σ_1. And Goldbach's conjecture that every even number greater than two is the sum of two primes can be expressed by the sentence

$$\forall x\{(\psi(x) \wedge 4 \leq x) \rightarrow (\exists y \leq x)(\exists z \leq x)(\chi(y) \wedge \chi(z) \wedge y + z = x)\}$$

which is (strictly) Π_1, since what is after the initial quantifier is built out of Δ_0 wffs using bounded quantifiers and connectives and so is still Δ_0.

Because (the formal expression of) Goldbach's conjecture is Π_1, such Π_1 sentences are often said to be of *Goldbach type*. True, Goldbach's conjecture is very simply expressed while other Π_1 sentences can be arbitrarily long and messy. But the thought is that all Π_1 sentences are like Goldbach's conjecture in involving, in effect, just universal quantification(s) of bounded stuff.

(d)   Next note the following simple result, to help fix ideas:

Theorem 24 *(i) The negation of a Δ_0 wff is Δ_0. (ii) The negation of a Σ_1 wff is a Π_1 wff, (iii) the negation of a Π_1 wff is a Σ_1 wff.*

Proof (i) holds by definition.

(ii) By definition, the negation of any Σ_1 wff is equivalent to some $\neg\varphi$ where φ is strictly Σ_1.

Working on the wff $\neg\varphi$, use the equivalences of '$\neg\exists\xi$' with '$\forall\xi\neg$', of '$\neg(\exists\xi \leq \kappa)$' with '$(\forall\xi \leq \kappa)\neg$', and of '$\neg(\forall\xi \leq \kappa)$' with '$(\exists\xi \leq \kappa)\neg$', together with de Morgan's laws, to 'drive negations inwards', until the negation signs all attach to the Δ_0 wffs that φ is ultimately built from. You will end up with a logically equivalent wff ψ built up using universal quantifiers, bounded quantifiers, conjunction and disjunction applied to those now negated Δ_0 wffs. But negated Δ_0 wffs are themselves still Δ_0 by (i). So $\neg\varphi$'s logical equivalent ψ is indeed strictly Π_1.

But what is equivalent to a strictly Π_1 wff is itself Π_1.

(iii) is established similarly.   ⊠

8.4   Q is Σ_1-complete

We now get to an easy but important result

Theorem 25 Q *is Σ_1-complete – i.e. it can prove any true Σ_1 sentence.*

Proof We need only look at the strict cases, since if Q can prove the true strict Σ_1 sentences, it can of course prove their logical equivalents too.

Define the degree of a Σ_1 sentence to be the number of applications of connectives and (bounded or unbounded) quantifiers needed to build it up from Δ_0 wffs. Then arrange the *true* strict Σ_1 sentences into some list, with lower degree sentences always preceding higher degree ones. Start marching along the list.

Theorem 23 tells us Q can prove the lowest degree true Σ_1 sentences which we meet at the beginning of the list, for these are just the true Δ_0 sentences.

Suppose that we have proceeded a certain way along our list of Σ_1 truths, and every sentence we have met so far is provable in Q. We'll now show the next sentence χ must be provable too. There are five cases to consider:

i. The true sentence χ is of the form $(\varphi \wedge \psi)$ with φ and ψ both Σ_1 sentences. Given that χ is true, φ and ψ must both be true sentences of lesser degree, so by assumption come earlier in the list and have already been seen to be provable in Q; and hence their conjunction is provable in Q.

ii. The case where the sentence χ is of the form $(\varphi \vee \psi)$ is dealt with similarly.

iii. The true sentence χ is of the form $(\forall \xi \leq \bar{n})\varphi(\xi)$. Since χ is true, so are $\varphi(0), \varphi(1), \ldots, \varphi(\bar{n})$. But these are true Σ_1 sentences of lesser degree, so by assumption come earlier in the list and must be provable in Q. So their conjunction is provable, and that means Q also proves χ by one of the results noted in §8.1.

iv. The case where χ is of the form $(\exists \xi \leq \bar{n})\varphi(\xi)$ is dealt with similarly.

v. The true sentence χ is of the form $\exists \xi \varphi(\xi)$. Since χ is true, some number n must satisfy φ, hence $\varphi(\bar{n})$ is true and of lesser degree, so comes earlier in our list. But then Q can prove $\varphi(\bar{n})$ by assumption, and hence can prove χ by existential quantifier introduction.

Hence, however χ is built up, it must be provable in Q.

Therefore, as we go down our list of truths, every sentence we meet must be provable in Q.   ⊠

8.5   A remarkable corollary

Our last theorem looks very straightforward and quite unremarkable; but it has a perhaps surprising corollary which is well worth noting:

Theorem 26 *If T is a consistent theory which includes Q, then every Π_1 sentence that it proves is true.*

Proof Suppose T proves a *false* Π_1 sentence φ. Then $\neg\varphi$ will be a *true* Σ_1 sentence. But in that case, since T includes Q and so is Σ_1-complete, T will also prove $\neg\varphi$, making T inconsistent. Contraposing, if T is consistent, any Π_1 sentence it proves is true. $\boxtimes$

Which is, in its way, a quite remarkable observation. It means that we don't have to fully *believe* a theory T – i.e. we don't have to accept that all its theorems are *true* on the interpretation built into T's language – in order to use it to establish that some Π_1 arithmetic generalization is true.

For example, with minor trickery, we can state Fermat's Last Theorem as a Π_1 sentence. And famously, Andrew Wiles has shown how to derive this Π_1 sentence from some *extremely* heavy-duty infinitary mathematics. Now we see, intriguingly, that this background mathematical theory does not need to be *sound* (have *true* axioms) – whatever exactly that means when things get so very wildly infinitary. It is enough for Wiles's proof successfully to establish that Fermat's Last Theorem is true that his background theory is *consistent*.

8.6 Intermediate arithmetics

We said at the beginning of the previous chapter that, in moving on from the very weak arithmetics BA and Q to consider first-order PA, we were jumping over a whole family of theories of intermediate strength. We can now briefly describe those intermediate theories: they are the ones we get by restricting the quantifier complexity of suitable instances of the induction schema.

For example, $I\Sigma_1$ is the theory we get by taking the first six axioms of PA (§7.5) plus the closure of every instance of the Induction Schema

$$\{\varphi(0) \,\wedge\, \forall x(\varphi(x) \to \varphi(Sx))\} \,\to\, \forall x\varphi(x),$$

where $\varphi(x)$ is now an open Σ_1 wff of L_A.

There is *technical* interest in knowing how much a theory like $I\Sigma_1$ can prove (as we will see in §20.1). But do such theories have any *conceptual* interest? After all, we gave reasons in §7.3 for being generous with induction: if $\varphi(x)$ expresses a genuine arithmetical property, how can induction fail for $\varphi(x)$?

Let's backtrack for a moment and ask again: which L_A open wffs $\varphi(x)$ *do* express genuine properties? Previously we took it that they all do: that is why we earlier suggested that any such wff $\varphi(x)$ is 'suitable' for appearing in an instance of the Induction Schema. But suppose you are a *very* stern constructivist who thinks that an arithmetical expression $\varphi(x)$ only really, *really*, makes sense if we can effectively decide whether or not it holds true of a given number. Or perhaps, slightly less sternly, you allow that the expression makes sense if it is Σ_1 so at least we can always prove it true of a given number when it is. *Then* you might want to restrict the induction principle to suitable instances using only Σ_1 expressions. But it would take us far too far afield even to begin to explore the possible merits of such stern proposals here.

Interlude

Let's pause to take stock. After the short biographical note in Chapter 1, we continued with:

Chapter 2. We met the First Incompleteness Theorem in this rough form: a nice enough theory T (which contains the language of basic arithmetic) will always be negation incomplete – there will always be sentences of basic arithmetic it can neither prove nor disprove.

Chapter 3. We then noted that we can cash out the idea of being a 'nice enough' theory in two ways. We can assume T to be *sound*. Or, retreating from that semantic assumption, we can require T to be (roughly) a *consistent theory which proves a modest amount of arithmetic*.

Chapter 4. Of course, we didn't *prove* the Theorem in either version, there at the very outset. However, in this next chapter, we waved an arm rather airily at the basic strategy that Gödel uses to establish the Theorem – namely we 'arithmetize syntax' (i.e. numerically code up facts about provability in ways that we can express in formal arithmetic) and then construct a Gödel sentence for a theory T that is true if and only if it isn't provable-in-T.

Chapter 5. We now did a bit better, in the sense that this chapter actually gave a *proof* that a consistent, effectively axiomatized, sufficiently strong, formal theory cannot be negation complete.

The argument was revealing, because it shows that we can get incompleteness results without calling on the arithmetization of syntax and the construction of Gödel sentences. However, the argument depends on the notion of 'sufficient strength' which is defined in terms of the informal notion of a 'decidable property' (a theory, remember, is sufficiently strong if it captures every decidable property of the natural numbers). And the discussion in Chapter 5 doesn't explain how we can sharpen up that informal notion of a decidable property, nor does it explain what a sufficiently strong theory might look like.

Chapter 6. We need to get less abstract, and start thinking about specific theories of arithmetic. So in this chapter, as a warm-up exercise, we first looked at BA, the quantifier-free arithmetic of the addition and multiplication of particular numbers. This is a negation-complete and decidable theory – but of course the theory is only complete, i.e. is only able to decide every sentence constructible in its language, because its quantifier-free language is so very limited. However, if we augment the language of BA by allowing ourselves the usual apparatus of first-order quantification, and replace the schematically presented axioms of BA with their obvious universally quantified correlates (and add in the axiom that every number bar zero is a successor) we get the much more interesting Robinson Arithmetic Q.

Since we are considerably enriching what can be expressed in our arithmetic language while not greatly increasing the power of our axioms, it is no surprise that Q is negation incomplete. And we can prove this without any fancy Gödelian considerations. We can easily show, for example, that Q can't prove either $\forall x(0 + x = x)$ or its negation. Q, then, is a very weak arithmetic.

Still, it will turn out to be the 'modest amount of arithmetic' needed to get a syntactic version of the First Theorem to fly. We announced (but of course haven't proved) that even Q is sufficiently strong: which explains why Q turns out to be so interesting despite its weakness.

Chapter 7. We then moved on to introduce first-order Peano Arithmetic PA, which adds to Q a whole suite of induction axioms (every instance of the Induction Schema). Exploration reveals that this theory, in contrast to Q, is very rich and powerful. We might, pre-Gödel, have very reasonably supposed that it is a negation-complete theory of the first-order arithmetic of addition and multiplication. But the theory is still effectively axiomatized, and the First Theorem is going to apply (assuming PA is sound, or is at least consistent and satisfies another syntactic condition). So PA too will turn out to be negation incomplete.

Chapter 8. There are theories intermediate in strength between Q and PA, theories which have induction axioms but only for wffs up to some degree of quantificational complexity. For technical reasons, we will later be interested in one such intermediate theory (see Chapter 20). But the task of Chapter 8 was just to explain this notion of quantificational complexity, and in particular explain what Σ_1 and Π_1 wffs are.

Which brings us up to the current point in this book. To give a sense of direction, let's now outline where we are going in the next five chapters. (Skip if you don't want spoilers!)

Chapter 9. The formal theories of arithmetic that we've looked at so far have (at most) the successor function, addition and multiplication built in. But why stop there? Even high-school arithmetic acknowledges many more numerical functions, like the factorial and the exponential.

In this chapter we describe a very wide class of such numerical functions, the so-called primitive recursive (p.r.) ones. They are a major subclass of the effectively computable functions.

We will also define the primitive recursive properties and relations – a numerical property/relation is p.r. when some p.r. function can effectively decide when it holds.

Chapter 10. Next, we show that L_A, the now-familiar formal language of basic arithmetic, can *express* all p.r. functions and relations.

Moreover Q and hence PA can *capture* all those functions and relations too – i.e. they can case-by-case prove wffs that assign the right values to the functions for particular numerical arguments. So Q and PA, despite having only successor, addition and multiplication 'built in', can actually deal with a vast range of functions (at least in so far as they can 'calculate' the value of the functions for arbitrary numerical inputs).

Note the link with our earlier talk about 'sufficiently strong theories' (cf. Defn. 20). Those, recall, are theories that can capture all effectively decidable properties of numbers. Well, now we are going to show that PA (indeed, even Q) can capture at least all those effectively decidable properties of numbers which are primitive recursive. And we'll find that that's enough for the core Gödelian argument to go through.

Chapter 11. We next re-introduce the key idea of the 'arithmetization of syntax' by Gödel-numbering, an idea which we first met in §§4.3 and 4.4. Focus on PA for the moment, and fix on a suitable Gödel-numbering. Then we can define various numerical properties/relations such as:

> $Wff(n)$ iff n is the code number of a PA-wff;
> $Sent(n)$ iff n is the code number of a PA-sentence;
> $Prf(m, n)$ iff m is the code number of a PA-proof of the sentence with code number n.

Moreover – the crucial result – these properties/relations are primitive recursive. Similar results obtain for any sensibly axiomatized formal theory.

Those last three chapters are inevitably rather action-packed (though I try to make the key Big Ideas as accessible as I can, and leave some details for enthusiasts to fill in from elsewhere). The effort, however, now really pays off:

Chapter 12. Since *Prf* is p.r., and the theory PA can express all p.r. relations, we can express some facts about PA-proofs in PA itself. We can now use this fact in constructing a Gödel sentence which is true if and only if it is not provable in PA. We can thereby prove a semantic version of Gödel's first incompleteness theorem for PA in something close to Gödel's way, assuming PA is sound. The result generalizes to other sensibly axiomatized sound arithmetics that extend the theory Q.

Chapter 13. Then in this chapter we at last prove a crucial syntactic version of the First Incompleteness Theorem, in something close to Gödel's own way.

So now read on again ...!

9 Primitive recursive functions

As just announced in the Interlude, the primitive recursive functions form a major subclass of the functions which are effectively computable (in the sense of Defn. 2). This chapter explains which functions these are, and proves some elementary results about them.

9.1 Introducing the primitive recursive functions

(a) Let's start by revisiting the basic axioms for addition and multiplication. We adopted formal versions in Q and PA. But now, throughout this chapter, we will be doing everyday informal mathematics. So – in keeping with our notational convention, §4.1 – here are the axioms re-presented informally, with the variables running over the natural numbers, but following the common practice of leaving quantifiers to be understood:

$$x + 0 = x$$
$$x + Sy = S(x + y)$$

$$x \times 0 = 0$$
$$x \times Sy = (x \times y) + x$$

The first of the pair of equations for addition tells us the result of adding zero to any number x. The second tells us the result of adding Sy to x in terms of the result of adding y. Hence these equations – as we pointed out in §6.2 – together tell us how to add any of $0, S0, SS0, SSS0, \ldots$, i.e. they tell us how to add *any* number to a given number x. Similarly, the first of the pair of equations for multiplication tells us the result of multiplying by zero. The second equation tells us the result of multiplying by Sy in terms of the result of multiplying by y and doing some addition. Hence the equations for multiplication and addition taken together tell us how to multiply a given number x by *any* number.

(b) Let's have two more very elementary examples. Take, then, the *factorial* function $y!$, which can be defined by the following two equations:

$$0! = 1$$
$$(Sy)! = y! \times Sy$$

The first equation assigns a value to the factorial function for the argument 0; the second equation tells us how to work out the value of the function for Sy

once we know its value for y (assuming we already know about multiplication). By laboriously applying and reapplying the second equation, we can successively calculate 1!, 2!, 3!, 4! ..., as follows:

$$1! = 0! \times 1 = 1$$
$$2! = 1! \times 2 = 2$$
$$3! = 2! \times 3 = 6$$
$$4! = 3! \times 4 = 24$$

And so on and on it goes. Our two-equation definition is indeed properly called a definition because it fixes the value of '$y!$' for all numbers y.

For our next example – this time another two-place function – consider the *exponential*. Using the standard notation, this can be defined by a similar pair of equations:

$$x^0 = S0$$
$$x^{Sy} = (x^y \times x)$$

The first equation gives the function's value for any value of x when $y = 0$, and – keeping x fixed – the second equation gives the function's value for the argument Sy in terms of its value for y (again assuming we already know about multiplication). The equations determine, e.g., that $3^4 = 3 \times 3 \times 3 \times 3 = 81$.

(c) Of course, you knew all that! But these very elementary examples are all we need in order to introduce the following key points:

  i. Note that, in each of these definitional pairs of equations, the second equation fixes the value of a function for argument Sy by invoking the value of the *same* function for argument y. A procedure where we evaluate a function for one input by calling the *same* function for a smaller input is standardly termed 'recursive' – and the particularly simple pattern we've illustrated is called, more precisely, 'primitive recursive'. So our two-clause definitions are examples of *definition by primitive recursion*.[1]

 ii. Next note, for example, that $(Sy)!$ is defined as $y! \times Sy$, so it is evaluated by evaluating $y!$ and Sy and then feeding the results of these computations into the multiplication function. This involves, in a word, the *composition* of functions, where evaluating a composite function involves taking the output(s) from one or more functions, and treating these as inputs to another function.

iii. Our four examples can then be arranged into two short *chains* of definitions by recursion and functional composition. Working from the bottom up, addition is defined in terms of the successor function; multiplication is then defined in terms of successor and addition; then the factorial (or, in

---

[1] "Surely, defining a function in terms of that very same function is circular!" But think of the relevant function as being defined in successive stages. The value of the function for input 0 is given. For each n in turn, the value of the function for Sn is then fixed by its already-settled value for n. So – by induction! – the function thereby does indeed get a determinate value for every numerical input.

the second chain, exponentiation) is defined in terms of multiplication and successor.

With these simple motivating examples in mind, here is our first, quick-and-dirty, way of specifying the p.r. functions:

A primitive recursive function is one that can be similarly characterized using a chain of definitions by primitive recursion and composition, starting from trivial 'initial functions' like the successor function.[2]

9.2   Defining the p.r. functions more carefully

We must now do better. (a) We need to tidy up the idea of defining a function by primitive recursion. (b) We need to tidy up the idea of defining a new function by composing old functions. And (c) we need to say more about the starter pack of initial functions which we can use in building up a chain of definitions by primitive recursion and/or composition.

(a)   Consider the recursive definition of the factorial again:

$$0! = 1$$
$$(Sy)! = y! \times Sy$$

This is an example of the following general scheme for defining a one-place function f:

$$f(0) = g$$
$$f(Sy) = h(y, f(y))$$

Here, g is just a number, while h is a two-place function *which we are assumed already to know about* prior to the definition of f. Maybe that is because h is an 'initial' function that we are allowed to take for granted. Or maybe it is because we've already given recursion clauses to define h. Or maybe h is a composite function constructed by plugging one known function into another – as in the case of the factorial, where $h(y, z) = z \times Sy$ (where we take the output from the successor function as one input into the multiplication function).

Likewise, with a bit of massaging, the recursive definitions of addition, multiplication and the exponential can all be treated as examples of the following general scheme for defining two-place functions:

$$f(x, 0) = g(x)$$
$$f(x, Sy) = h(x, y, f(x, y))$$

---

[2]History: the basic idea is there in a classic paper by Richard Dedekind in 1888, and highlighted again by Thoralf Skolem in 1923. But the modern terminology 'primitive recursion' seems to be due to Rósza Péter in 1934; and 'primitive recursive function' was first used by Stephen Kleene in 1936.

I'll use the abbreviation 'p.r.': but note that the same abbreviation is quite often used elsewhere as short for '*partial* recursive' – a quite different notion!

where now g is a one-place function, and h is a three-place function, both of them functions that we already know about. Three points about this:

i. To get the definition of addition to fit this pattern, with a unary function on the right of the first equation, we take $g(x)$ to be the trivial *identity function* $I(x) = x$.

ii. To get the definition of multiplication to fit the pattern, $g(x)$ has to be treated as the equally trivial *zero function* $Z(x) = 0$.

iii. How do we get the definition of addition $x + Sy = S(x + y)$ to fit the pattern $f(x, Sy) = h(x, y, f(x, y))$?

We have to take $h(x, y, z)$ to be the function Sz. So, as this illustrates, we must allow h not to care what happens to some of its arguments, while operating on some other argument(s). The conventional way of doing this is to help ourselves to some further trivial identity functions that serve to select out particular arguments. For example, the function I_3^3 takes three arguments, and just returns the third of them, so $I_3^3(x, y, z) = z$. Then, in the definition of addition, we can put $h(x, y, z) = SI_3^3(x, y, z)$, so h is defined by composition from initial functions which we can take for granted.

We can now generalize the idea of a definition by recursion from the case of one-place and two-place functions to cover the case of many-place functions. There's a standard notational device that helps to put things snappily: we write $\vec{x}$ as short for the array of k variables $x_1, x_2, \ldots, x_k$ (taking the relevant k to be fixed by context). Then we can say:

Defn. 26 *Suppose that the following holds:*

$$f(\vec{x}, 0) = g(\vec{x})$$
$$f(\vec{x}, Sy) = h(\vec{x}, y, f(\vec{x}, y))$$

Then f is defined from g and h by primitive recursion.

This covers the case of one-place functions $f(y)$ like the factorial if we allow $\vec{x}$ to be empty, in which case $g(\vec{x})$ is a 'zero-place function', i.e. a constant.

(b)   Now to tidy up the idea of definition by composition. The basic idea, to repeat, is that we form a composite function f by treating the output value(s) of one or more given functions $g, g', g'', \ldots$, as the input argument(s) to another function h. For example, we set $f(x) = h(g(x))$. Or, to take a slightly more complex case, we could set $f(x, y, z) = h(g(x, y), g'(y, z))$.

There's a number of equivalent ways of covering the manifold possibilities of composing multi-place functions. But one simple way is to define what we might call one-at-a-time composition (where we just plug *one* function g into another function h), thus:

Defn. 27 *If $g(\vec{y})$ and $h(\vec{x}, u, \vec{z})$ are functions – with $\vec{x}$ and $\vec{z}$ possibly empty – then f is defined by composition by substituting g into h just if $f(\vec{x}, \vec{y}, \vec{z}) = h(\vec{x}, g(\vec{y}), \vec{z})$.*

61

We can then think of generalized composition – where we plug more than one function into another function – as just iterated one-at-a-time composition. For example, we can substitute the function $g(x, y)$ into $h(u, v)$ to define the function $h(g(x, y), v)$ by composition. Then we can substitute $g'(y, z)$ into the defined function $h(g(x, y), v)$ to get the composite function $h(g(x, y), g'(y, z))$.

(c)   Now, our quick-and-dirty characterization of the primitive recursive functions said that they are built up by recursion and composition, beginning from some 'starter pack' of trivial basic functions.

But which functions will count as basic? We have in fact met all we need:

Defn. 28 *The initial functions are the successor function S, the zero function $Z(x) = 0$ and all the k-place identity functions, $I_i^k(x_1, x_2, \ldots, x_k) = x_i$ for each k, and for each i, $1 \leq i \leq k$.*

These identity functions are also often called *projection* functions (compare projecting the vector with components $x_1, x_2, \ldots, x_k$ onto the i-th axis).

Given (a) to (c), we can now put everything together, and give the following official account:

Defn. 29 *The p.r. functions are as follows:*

(1) *The successor function S, zero function Z, and all the identity functions I_i^k are p.r.;*
(2) *if f can be defined from the p.r. functions g and h by composition, substituting g into h, then f is p.r.;*
(3) *if f can be defined from the p.r. functions g and h by primitive recursion, then f is p.r.;*
(4) *nothing else is a p.r. function.*

(We allow g in clauses (2) and (3) to be zero-place, i.e. be a constant.)

A p.r. function f is therefore one that *can* be specified by a chain of definitions by recursion and composition, leading back to initial functions. Which accords with the informal characterization at the end of the last section.

Let's have another mini-example of such a chain, this time to show that the absolute difference function is primitive recursive. So consider:

$$P(0) = 0$$
$$P(Sx) = x$$

$$x \mathbin{\dot-} 0 = x$$
$$x \mathbin{\dot-} Sy = P(x \mathbin{\dot-} y)$$

$$|x - y| = (x \mathbin{\dot-} y) + (y \mathbin{\dot-} x)$$

Here, 'P' signifies the predecessor function (with zero being treated as its own predecessor), and its two-part definition is a trivial case of a definition by primitive recursion. Next, '$\mathbin{\dot-}$' signifies 'subtraction with cut-off', so $m \mathbin{\dot-} n$ is zero if $m < n$. Then $|m - n|$ is the absolute difference between m and n.

9.3 How to prove a result about all p.r. functions

Let's introduce an obvious bit of terminology:

Defn. 30 *A definition chain for the p.r. function f is a sequence of functions $f_0, f_1, f_2, \ldots, f_k$ where each f_j is either an initial function or is defined from previous functions in the sequence by composition or primitive recursion, and $f_k = f$.*

The closure condition (4) in the previous definition means that every p.r. function is required to have a definition chain in this sense (the chain need not be unique, but a function must have at least one to be p.r.). Which means in turn that there is a simple method of proving that every p.r. function shares some feature. For suppose that, for some given property P, we can show the following:

P1. The initial functions have property P.

P2. If the functions g and h have property P, and f is defined by composition from g and h, then f also has property P.

P3. If the functions g and h have property P, and f is defined by primitive recursion from g and h, then f also has property P.

It then follows that all primitive recursive functions have property P.

  Why? Take any p.r. function f. It must have a definition chain. Now trek along such a chain for f. Each initial function we encounter has property P by P1. By P2 and P3, each definition by recursion or composition which is used in the chain takes us from functions which have property P to another function with property P. So, every function we define as we go along the chain has property P, including the final target function f.

  In sum, then: to prove that all p.r. functions have some property P, it suffices to prove the relevant versions of P1, P2 and P3.

  For a very simple (but important!) first example, take the property of being a *total function* of the natural numbers, i.e. being a function which successfully outputs a natural number value for any given numerical input. The initial functions are, trivially, total functions of numbers, defined for every numerical argument; also, primitive recursion and composition both build total functions out of total functions (why? check this claim!). Which means that p.r. functions are always total functions, defined for all natural number arguments.

9.4 The p.r. functions are computable

We now show that every p.r. function is effectively computable in the sense of Defn. 2. And we'll take the discussion in stages.

(a)   Given the general strategy just described, it is enough to show:

C1. The initial functions are effectively computable.

C2. If f is defined by composition from effectively computable functions g and h, then f is also effectively computable.

63

C3. If f is defined by primitive recursion from the effectively computable functions g and h, then f is also effectively computable.

For C1 it is enough to remark that the trivial initial functions S, Z, and I_i^k count as computable if any do!

For C2 simply note that to compute the composition of two computable functions g and h you just feed the output from whatever algorithmic routine evaluates g as input into the routine that evaluates h.

To illustrate C3, return once more to our example of the factorial. Here is its p.r. definition again,

$$0! = 1$$
$$(Sy)! = y! \times Sy$$

The first clause gives the value of the function for the argument 0; then – as we said – you can repeatedly use the second recursion clause to calculate the function's value for $S0$, then for $SS0$, $SSS0$, etc. So the definition in fact encapsulates an algorithm for effectively calculating the function's value for each successive number n (given that we already know how to compute multiplications).

This last observation evidently generalizes to other cases where we define a function by primitive recursion from known computable functions; which establishes C3.

(b) But there's more to be said.

Let's think again about the *kind* of algorithm needed to compute $n!$ (assuming that we can already handle multiplication). We need a computational routine which takes n as input, and proceeds like this:

1. Set the variable *fact* to the initial value 1.
2. Set the counter y to the initial value 0 and enter a loop.
3. If y equals n then exit the loop, else:
4.   Compute *fact* $\times Sy$, and make the result the new value of *fact*.
5.   Increment the value of the counter y by 1.
6.   Go back to (3).

The routine terminates when the counter reaches n with the variable *fact* now having the value $n!$.

And note: the crucial thing about executing this kind of loop is that the total number of iterations to be run through is fixed in advance of entering the loop (or at least, a maximum bound is set in advance, if we allow for early exits). It will be useful to have a label for such bounded-in-advance looping structures: let's call them *basic 'for' loops*.[3]

Now, our mini-program for the factorial calls the multiplication function which can itself be computed by a similar basic 'for' loop (invoking addition). And addition can be computed by another such 'for' loop (invoking the successor function). So reflecting the downward chain of recursive definitions

---

[3] Two reasons. Such loops are indeed a basic computational structure. And, fine details apart, they are instantiated by loops introduced by the keyword 'for' in some ancient programming languages like Basic.

factorial $\Rightarrow$ multiplication $\Rightarrow$ addition $\Rightarrow$ successor

there will be a composite program for the factorial containing *nested* basic 'for' loops, which ultimately calls primitive operations like incrementing the content of a variable by 1 or setting a variable to zero.

This point again generalizes, giving us the following key observation:

Theorem 27 *A primitive recursive function f is effectively computable by a program whose iterative constructions are all basic 'for' loops (possibly nested to a certain fixed depth, depending on the number of definitions by recursion in our definition chain for f).*

(c)   The converse of the last theorem is also true. Suppose a program sets a value g for $f(0)$, and then goes into a basic 'for' loop which which calls on some already computable function h which is used on loop number y (counting from zero) to fix the value of $f(Sy)$ in terms of the values of y and $f(y)$. This plainly corresponds to a definition f from g and h by primitive recursion. And the point also generalizes:

Theorem 28 *If a function can be computed by a program without open-ended searches – using just basic 'for' loops for iterative procedures (nested to a certain fixed depth) and with the program's 'built in' functions all being p.r. – then the newly defined function will also be primitive recursive.*

This gives us a quick way of convincing ourselves that a new function is p.r.: sketch out a routine for computing it and check that the needed looping computations only invoke already known p.r. functions and the number of iterations inside any loop is always bounded in advance, so we do not need to set off on any open-ended searches. Then the new function will be primitive recursive.

For a quick example, take the two-place function $gcd(x, y)$ which outputs the greatest common divisor of the two inputs. Evidently, a bounded search through cases is enough to do the trick: at its crudest and most inefficient, we can look in turn at all the numbers up to and including the smaller of x and y and see if it divides both. That sketched algorithmic procedure is enough to assure us that $gcd(x, y)$ is p.r. without going through the palaver of actually specifying a suitable definition chain.

9.5   Not all computable numerical functions are p.r.

(a)   The values of a given primitive recursive function can be computed by a program involving basic 'for' loops as its main programming structure. Here, each loop goes through a specified number of iterations, set in advance.

But now, for a contrasting case, recall our proof of Theorem 7 about the decidability of negation-complete, effectively decidable, theories. There, we allowed the process *enumerate the theorems and wait to see which of φ or $\neg\varphi$ turns up* to count as an effective computational decision procedure. In other words, we set out on an open-ended iterative procedure where there is no prior bound given

65

on the number of times we need to go around the block on our search.

Generalizing, we allow for computations which involve 'until' loops, where a routine can be repeatedly iterated until some condition is fulfilled, as many times as it takes, with no prior bound set.[4]

(b)   Given we do also permit procedures involving open-ended searches to count as effective computations, it looks very plausible that not everything which is effectively computable need be primitive recursive.

And in fact we can do better than offer plausibility considerations. We will now prove:

Theorem 29 *There are effectively computable numerical functions which aren't primitive recursive.*

Proof The p.r. functions are effectively enumerable. That is to say, there is an effective way of numbering off functions f_0, f_1, f_2, ..., such that each of the f_i is p.r., and each p.r. function appears somewhere on the list.

Why? A p.r. function is defined by recursion or composition from other functions which are defined by recursion or composition from other functions which are defined ... ultimately in terms of some primitive starter functions. So, as we put it before, every p.r. function has a definition chain. Choose some economical formal language for specifying such chains. Then we can effectively generate all possible strings of symbols from this specification language (arranged by length, then 'in alphabetical order'); and as we go along, we can effectively select out the strings that obey the rules for determining a p.r. function by specifying a definition chain. This generates a list which effectively enumerates the p.r. functions, $f_0, f_1, f_2, f_3, \ldots$, repetitions allowed.

So consider the following table:

	0	1	2	3	...
f_0	$\underline{f_0(0)}$	$f_0(1)$	$f_0(2)$	$f_0(3)$	...
f_1	$f_1(0)$	$\underline{f_1(1)}$	$f_1(2)$	$f_1(3)$	...
f_2	$f_2(0)$	$f_2(1)$	$\underline{f_2(2)}$	$f_2(3)$	...
f_3	$f_3(0)$	$f_3(1)$	$f_3(2)$	$\underline{f_3(3)}$	...
...	...	...	...	...	↘

---

[4]It would be misleading to baldly contrast 'for' loops with 'until' (or 'while') loops. For a start, a basic 'for' loop can be thought of as a special kind of 'until' loop where we increment a loop-counter on each iteration and continue for a fixed number of cycles round the loop until the counter reaches its given target. And conversely, in many computer languages, loops introduced by the keyword 'for' can be open-ended.

What matters, then, is not the particular keywords used, but – to re-emphasize – the difference between looping structures where the number of iterations is given a fixed numerical bound in advance, and those where it isn't.

Down the table we list off the p.r. functions. An individual row then gives the values of a particular f_n for each successive argument. Let's now define the corresponding *diagonal* function, by putting $\delta(n) = f_n(n) + 1$. To compute $\delta(n)$, we just run our effective enumeration of the definition chains for p.r. functions until we get to the recipe for f_n. We follow the instructions in that recipe to evaluate that function for the argument n. We then add one. Each step is entirely mechanical. So our diagonal function is effectively computable, using a step-by-step algorithmic procedure.

By construction, however, the function δ can't be primitive recursive. For suppose otherwise. Then δ must appear somewhere in the enumeration of p.r. functions, i.e. be the function f_d for some index number d. But now ask what the value of $\delta(d)$ is. By hypothesis, the function δ is none other than the function f_d, so $\delta(d) = f_d(d)$. But by the initial definition of the diagonal function, $\delta(d) = f_d(d) + 1$. Contradiction.

So we have, as they say, 'diagonalized out' of the class of p.r. functions to define a function δ which is still effectively computable but not primitive recursive. $\boxtimes$

(c)   "But hold on! *Why* is δ not a p.r. function?" Because, on the one hand, any given p.r. function will have a certain number of recursions in its definition, and so can be computed by a program with its looping procedures nested to some fixed depth. On the other hand, to evaluate δ for increasing values of n, we will have to evaluate the different functions f_n for argument n. Evaluating these different functions will call on computations involving loops nested to varying different depths. And we ordered the specifications of f_n in part by length, allowing for more and more recursions in their definitions, so the depth of these nested loops can tend to grow and grow without limit as n increases. So, unlike with a p.r. function, there is no fixed depth to the loops involved in computing the function δ.

9.6   Defining p.r. properties and relations

We have defined the class of p.r. *functions*. Finally in this chapter, we extend the scope of the idea of primitive recursiveness and introduce the ideas of *p.r. decidable (numerical) properties* and *relations*.

Now, quite generally, we can tie together talk of functions and talk of properties and relations by using the notion of a *characteristic function* which we touched on right back in §2.1:

Defn. 31 *The characteristic function of the numerical property P is the one-place function c_P such that if m is P, then $c_P(m) = 1$, and if m isn't P, then $c_P(m) = 0$.*

The characteristic function of the two-place numerical relation R is the two-place function c_R such that if m is R to n, then $c_R(m, n) = 1$, and if m isn't R to n, then $c_R(m, n) = 0$.

And similarly for many-place relations. The choice of values for the characteristic function is, of course, entirely arbitrary: any pair of distinct numbers would do. Our choice is supposed to be reminiscent of the familiar use of 1 and 0, one way round or the other, to stand in for *true* and *false*.[5]

The numerical property P partitions the numbers into two sets, the set of numbers that have the property and the set of numbers that don't. Its corresponding characteristic function c_P also partitions the numbers into two sets, the set of numbers the function maps to the value 1, and the set of numbers the function maps to the value 0. And these are the *same* partition. So in a good sense, P and its characteristic function c_P contain exactly the same information about a partition of the numbers: hence we can move between talk of a property and talk of its characteristic function without loss of information. Similarly, of course, for relations (which partition pairs of numbers, etc.). And we can use this link between properties and relations and their characteristic functions in order to carry over ideas defined for functions and apply them to properties/relations.

For example, without further ado, we now extend the idea of primitive recursiveness to cover properties and relations:

Defn. 32 *A p.r. decidable property is a property with a p.r. characteristic function, and likewise a p.r. decidable relation is a relation with a p.r. characteristic function.*

By way of casual abbreviation, we'll fall into saying that p.r. decidable properties and relations are themselves (simply) p.r.

For a quick example, consider the property of being a *prime* number. Take the characteristic function $pr(n)$ which has the value 1 when n is prime, and 0 otherwise. Now just note that we can evidently compute $pr(n)$ just using 'for' loops: we just do a bounded search through numbers less than n – indeed, no greater than $\sqrt{n}$ – and if we find a divisor of n other than 1, return the value 0, and otherwise return the value 1. So the property of being prime is p.r.

---

[5] *IGT*, for technical reasons, follows Gödel in using 0 rather than 1 for *true*.

10 Expressing and capturing the primitive recursive functions

We have reached the following point. In Chapters 6 and 7, we introduced some *formal* arithmetics with just three functions built in – successor, addition, multiplication. But in Chapter 9 we reminded ourselves that ordinary *informal* arithmetic talks about other elementary (and computable) functions such as the factorial, the exponential, and so on. And, reflecting on the way these functions are defined, we generalized to specify a whole class of primitive recursive functions (and corresponding p.r. properties and relations).

It might seem, then, that a gulf has opened up between the apparent modesty of the resources of our formal theories of arithmetic and the great richness of the world of p.r. functions. But not so. I've already announced in §6.10 that – despite its great limitations – even the weak theory Q can in fact capture all the p.r. properties and relations. This chapter explains how this is proved.

10.1 Two key theorems

We are going to first meet two pivotal results. Theorem 31 below will tell us about the expressive power of Q's language:

The language L_A can express all primitive recursive functions.

So, to consider just the simplest case, take a one-place p.r. function f. Then there will be a corresponding L_A-wff $\varphi(x, y)$ such that for any particular m, n, $\varphi(\overline{m}, \overline{n})$ is true if and only if $f(m) = n$. And there's more. Looking at the proof of this result we find that in fact L_A can express any p.r. function using a wff of low quantifier complexity: a Σ_1 wff suffices.

The other key theorem we establish will be Theorem 33, a corresponding claim about what Q can case-by-case prove:

The theory Q can capture any p.r. function by a Σ_1 wff.

We haven't yet officially defined 'capturing' for functions as opposed to relations (compare Defn 19). But the idea is predictable. The key requirement is that, given for example a one-place p.r. function f, there will be a corresponding Σ_1-wff $\varphi(x, y)$ such that, case by case, if $f(m) = n$, then Q proves $\varphi(\overline{m}, \overline{n})$, while if $f(m) \neq n$, then Q proves $\neg\varphi(\overline{m}, \overline{n})$.

Now, as we will see in this chapter, the main *ideas* involved in proving these pivotal theorems are rather neat and not at all difficult. But there's no getting away from it: working through all the proof-*details* would take a bit of patience. So I try to steer a middle way. I will give most of a proof of Theorem 31; but I will only outline in broad strokes what it takes to prove Theorem 33 and its corollary about Q's capturing p.r. properties and relations.

10.2 L_A can express the factorial function

In this section, then, we are going to look at a simple example and show that L_A – despite having only successor, addition and multiplication built in – can in fact *express* the factorial function. So we are aiming to construct an L_A wff $F(x, y)$ such that, for any particular m and n, $F(\overline{m}, \overline{n})$ if and only if $n = m!$.

In the next section we'll use the same key construction in showing that L_A can express *any* p.r. function at all. But it will be much easier to follow the general argument if you first meet the 'β-function trick' here, when deployed in our simple special case of the factorial. Even so, the details are still going to be a bit fiddly: we'll take things in five stages.

(a) Consider, then, the p.r. definition of the factorial function again:

$$0! = 1$$
$$(Sx)! = x! \times Sx$$

We can think of this definition in the following way: for any x, it tells us how to construct a sequence of numbers $0!, 1!, 2!, \ldots, x!$, where we move from the i-th member of the sequence (counting from zero[1]) to the next by multiplying by Si. Or putting it a bit more abstractly, suppose that for numbers x and y,

(1) There is a sequence of numbers $k_0, k_1, k_2, \ldots, k_x$ such that: $k_0 = 1$, and if $i < x$ then $k_{Si} = k_i \times Si$, and $k_x = y$.

Then this is equivalent to saying that $y = x!$.

So the question of how to reflect the p.r. definition of the factorial inside L_A can be parlayed into the following question: how can we express facts about *finite sequences of numbers* using the limited resources of L_A?

(b) By using numerical codes! Suppose that we can encode a finite sequence into a single *code number* c, and suppose we have a corresponding two-place *decoding function* – call it simply *decode* – such that if you give the function *decode* the code c and the index i, the function spits out the i-th member of the sequence which c codes. In other words, suppose that, when c is the code number for the sequence $k_0, k_1, k_2, \ldots, k_x$, then $decode(c, i) = k_i$.

If we can find such a coding scheme, then we can rewrite (1) as follows, by talking about a code number c instead of the sequence $k_0, k_1, k_2, \ldots, k_x$, and by writing $decode(c, i)$ instead of k_i:

---

[1] Note then, just for local convenience in *this* section, we number off items in a sequence starting from the zero-th.

(2) There is a code number c such that: $decode(c, 0) = 1$, and if $i < x$ then $decode(c, Si) = decode(c, i) \times Si$, and $decode(c, x) = y$.

This way, if a suitable *decode* function can indeed be expressed in L_A, then we can say $y = x!$ in L_A. Great! So can we do this coding trick?

(c) There's certainly nothing mysterious about the idea of a coding/decoding scheme for finite sequences of numbers. Informally, for example, we could use the rather obvious trick of coding the sequence $k_0, k_1, k_2, \ldots, k_x$ by using powers of primes, i.e. by setting the code number $c = 2^{k_0} \cdot 3^{k_1} \cdot 5^{k_2} \cdot \ldots \cdot \pi_x^{k_x}$ (where π_x is the x-th prime, counting from 0). Then the corresponding $decode(c, i)$ would output the exponent of the ith prime factor of c. The uniqueness of prime factorization ensures that this coding/decoding scheme works as desired.

Unfortunately, however, we can't use *this* coding device inside L_A since our formal language doesn't have exponentiation built in. We are instead going to use a neat, but less obvious, trick due to Gödel.[2]

As a first step, let's liberalize our idea of a coding/decoding scheme just a little. We will now aim to code a sequence using *two* numbers c and d, and then our corresponding decoding function will take *two* code numbers c and d and an index number i. So:

> A two-code scheme for coding finite sequences provides a function $decode(c, d, i)$ such that, for *any* finite sequence of natural numbers $k_0, k_1, k_2, \ldots, k_x$ there is a *pair* of numbers c, d such that, for every $i \leq x$, $decode(c, d, i) = k_i$.

But even with this liberalization, it still isn't obvious how to define a decoding function in terms of the functions built into basic arithmetic. But Gödel solved the problem with his β-function. Put

$\beta(c, d, i) =_{\text{def}}$ the remainder left when c is divided by $d(i + 1) + 1$.

Then we have, just as we want,

Theorem 30 *For any finite sequence of numbers $k_0, k_1, \ldots, k_n$, we can find a suitable pair of numbers c, d such that for $i \leq n$, $\beta(c, d, i) = k_i$.*

This arithmetical claim should look intrinsically plausible. As we divide c by $d(i + 1) + 1$, then for different values of i ($0 \leq i \leq n$) we'll get a sequence of $n + 1$ remainders. Vary c and d, and the sequence of remainders will vary. The permutations as we vary c and d without limit *appear* to be simply endless. We just need to check, then, that appearances don't deceive, and we *can* always

---

[2]Fine print. Arguably, it would have been more elegant to start by considering a language L_A^* which *does* have exponentiation built in, and correspondingly start from augmented arithmetics Q* and PA* which have built-in axioms for exponentiation. We could then have more easily proved that L_A^* expresses and that Q* (and so PA*) captures all p.r. functions, and gone on to prove corresponding incompleteness theorems.

Afterwards, as a footnote so to speak, we could *then* have shown that by Gödel's trick we get the same results even if exponentiation isn't built it. But we are following the more usual path of starting with unaugmented L_A, and unaugmented Q/PA, and so now have to tangle with the β-function manoeuvre upfront.

71

find a (big enough) c and a (smaller) d which makes the sequence of remainders match a given $n + 1$-term sequence of numbers.[3]

(d)   We said that $y = x!$ just in case

(1) There is a sequence of numbers $k_0, k_1, \ldots, k_x$ such that: $k_0 = 1$, and if $i < x$ then $k_{Si} = k_i \times Si$, and $k_x = y$.

And we now know, thanks to Gödel, that we can reformulate this by using his β-function as follows:

(2′) There is some pair of code numbers c, d such that: $\beta(c, d, 0) = 1$, and if $i < x$ then $\beta(c, d, Si) = \beta(c, d, i) \times Si$, and $\beta(c, d, x) = y$.

But the β-function is defined in terms of the elementary arithmetical notion of remainder-on-division. So we should expect that it can be expressed in L_A by some open wff which we will abbreviate B – meaning, of course, that for particular numbers a, b, j, k, $\mathsf{B}(\overline{a}, \overline{b}, \overline{j}, \overline{k})$ is true if and only if $\beta(a, b, j) = k$.

Assuming for a moment that a suitable wff B really is available, (2′) can then be rendered into L_A as follows:

(3) $\exists c \exists d \{ \mathsf{B}(c, d, 0, S0) \land$
    $(\forall i \leq x)[i \neq x \rightarrow \exists v \exists w \{ (\mathsf{B}(c, d, i, v) \land \mathsf{B}(c, d, Si, w)) \land w = v \times Si \}] \land$
    $\mathsf{B}(c, d, x, y) \}.$

Only the variables 'x' and 'y' remain free, and we can abbreviate that wff by 'F(x, y)'. Since (3) renders something equivalent to (1), it follows that – as wanted – F(x, y) expresses the factorial function.

(e)   Which is all really neat! – assuming we can indeed come up with an L_A-wff B which expresses β. But just reflect that the concept of a remainder on division can be elementarily defined in terms of multiplication and addition. The remainder when a is divided by b equals y just when there is some number u (no greater than a) such that $a = (b \times u) + y$, where $y < b$.

Similarly, the remainder left when c is divided by $d(i+1)+1$ equals y just when there is some number u (no greater than c) such that $c = ((d(i+1)+1) \times u) + y$, where $y < d(i+1) + 1$, or equivalently $y \leq d(i+1)$.

So consider the following open wff (with its initial bounded quantifier):

$\hat{\mathsf{B}}(c, d, i, y) =_{\text{def}} (\exists u \leq c)[c = (S(d \times Si) \times u) + y \land y \leq (d \times Si)].$

This, as we wanted, expresses our three-place Gödelian β-function in L_A (for remember, we can define bounded quantifiers in L_A).

Job done. Except that, for future technical reasons, it turns out to be useful to add to $\hat{\mathsf{B}}$ a clause that reflects that the β function is indeed a function (and hence, in particular, when it sends inputs (c, d, i) to the value y, it can't send those same inputs to any smaller value z). So let's define

$\mathsf{B}(c, d, i, y) =_{\text{def}} \hat{\mathsf{B}}(c, d, i, y) \land (\forall z \leq y)(z \neq y \rightarrow \neg \hat{\mathsf{B}}(c, d, i, z)).$

This then will be our official way of expressing the β function in L_A.

---

[3]Mathematical completists: see *IGT2*, §15.2, fn. 4 for a proof that this works!

10.3 L_A can express all p.r. functions

That was rather fiddly. But the Big Ideas governing the construction were straightforward enough (I hope!). And what goes for the factorial function goes across the board. Using the β-function trick again to handle any definitions by recursion, we now show that L_A can express any p.r. function.

You are welcome to take this generalizing claim largely on trust and skip quite lightly over the outline proof. But don't skip entirely, as the proof-sketch introduces an important new idea we'll need later.

So to continue. We already know the standard strategy for showing that something is true of all p.r. functions. Suppose then that we establish:

E1. L_A can express the initial functions.

E2. If L_A can express the functions g and h, then it can also express a function f defined by composition from g and h.

E3. If L_A can express the functions g and h, then it can also express a function f defined by primitive recursion from g and h.

Then by the argument of §9.3, those assumptions will be enough to establish our desired general result. So how can we prove (E1) to (E3)?

(E1) will be trivial. And (E2) looks easy enough – just suitably combine the wffs which express g and h to get a wff which expresses the composition of those functions. (E3) takes a bit more work – but you now know the trick to use: wheel out the β-function again! In more detail:

Proof (of E1) Just look at cases. The successor function $Sx = y$ is of course expressed by the wff $Sx = y$.

The zero function, $Z(x) = 0$ is expressed by the wff $Z(x, y) =_{\text{def}} (x = x \wedge y = 0)$.

Finally, the three-place function $I_2^3(x, y, z) = y$, to take just one example of an identity function, is expressed by the wff $I_2^3(x, y, z, u) =_{\text{def}} y = u$ (or we could use $(x = x \wedge y = u \wedge z = z)$ if we'd like 'x' and 'z' actually to appear in the wff). Likewise for all the other identity functions.   ⊠

Proof (of E2) Suppose, to take a simple example, that g and h are one-place functions, expressed by the wffs $G(x, y)$ and $H(x, y)$ respectively. Then, the function $f(x) = h(g(x))$ is evidently expressed by the wff $\exists z(G(x, z) \wedge H(z, y))$ – i.e. for any m, n, $\exists z(G(\overline{m}, z) \wedge H(z, \overline{n}))$ if and only if $f(m) = n$. Why so?

Suppose $g(m) = k$ and $h(k) = n$, so $f(m) = n$. Then by hypothesis $G(\overline{m}, \overline{k})$ and $H(\overline{k}, \overline{n})$ will be true, and hence $\exists z(G(\overline{m}, z) \wedge H(z, \overline{n}))$ is true, as required.

Conversely, suppose $\exists z(G(\overline{m}, z) \wedge H(z, \overline{n}))$ is true. Then since the quantifiers run over numbers, $(G(\overline{m}, \overline{k}) \wedge H(\overline{k}, \overline{n}))$ must be true for some k. So we'll have $g(m) = k$ and $h(k) = n$, and hence $f(m) = h(g(m)) = n$ as required.

Other cases of composition can be handled similarly.   ⊠

Proof (of E3) We need to show that we can use the β-function construction again and – exactly following the model of our treatment of the factorial – prove more generally that, if the function f is defined by recursion from functions g

and h which are already expressible in L_A, then f is also expressible in L_A.

We are assuming that

$$f(\vec{x}, 0) = g(\vec{x})$$
$$f(\vec{x}, Sy) = h(\vec{x}, y, f(\vec{x}, y)).$$

(Remember, $\vec{x}$ indicates variables being 'carried along for the ride', that don't change in the course of the recursion that defines $f(\vec{x}, Sy)$ in terms of $f(\vec{x}, y)$.)

This definition tells us that $f(\vec{x}, y) = z$ just in case:

(1) There is a sequence of numbers $k_0, k_1, \ldots, k_y$ such that: $k_0 = g(\vec{x})$, and if $i < y$ then $k_{Si} = h(\vec{x}, i, k_i)$, and $k_y = z$.

Equivalently, using our β-function again, $f(\vec{x}, y) = z$ is true when:

(2) There is some c, d, such that: $\beta(c, d, 0) = g(\vec{x})$, and if $i < y$ then $\beta(c, d, Si) = h(\vec{x}, i, \beta(c, d, i))$, and $\beta(c, d, y) = z$.

Assume there are n variables wrapped up in $\vec{x}$. And suppose we can already express the n-place function g by a $(n+1)$-variable expression G, and the $(n+2)$-variable function h by the $(n+3)$-variable expression H. Then – using '$\vec{x}$' to indicate a suitable sequence of n variables – (2) can be rendered into L_A by

(3) $\exists c \exists d \{ \exists k [B(c, d, 0, k) \land G(\vec{x}, k)] \land$
$\qquad (\forall i \leq y)[i \neq y \to \exists v \exists w \{ (B(c, d, i, v) \land B(c, d, Si, w)) \land H(\vec{x}, i, v, w) \}] \land$
$\qquad B(c, d, y, z) \}.$

Abbreviate this defined wff as $\varphi(\vec{x}, y, z)$; it is then evident that φ will serve to express the p.r. defined function f. Which gives us the desired result E3. ⊠

So, we've shown how to establish each of the claims E1, E2 and E3. Hence

Theorem 31 *The language L_A can express all primitive recursive functions.*

Proof For any p.r. function f, there is a sequence of functions $f_0, f_1, f_2, \ldots, f_k$ where each f_j is either an initial function or is constructed out of previous functions by composition or recursion, and $f_k = f$.

Corresponding to that sequence of functions we can write down a sequence of L_A wffs which express each of those functions in turn. We write down the E1 expression corresponding to an initial function. If f_j comes from two previous functions in the sequence by composition, we use an existential quantifier construction as in E2 to write down a wff built out of the wffs expressing the two previous functions. And if f_j comes from two of the previous functions by recursion, we use the β-function trick and write down a (3)-style expression built out of the wffs expressing the two previous functions.

So this way we eventually build up to the wff which expresses f_k, i.e. our target function f. (Laborious, but it works!) ⊠

10.4 Canonical wffs for expressing p.r. functions are Σ_1

Now we extract from the sketched proof of our last theorem an important new idea, which we'll need later. So let's say that

Defn. 33 *An L_A wff canonically expresses the p.r. function f if it recapitulates a definitional chain for f by being constructed in the manner described in the proof of Theorem 31.*

We can express a given p.r. function f by other wffs too (for a start, by adding redundant clauses): but it is the *canonical* ones from which we can directly read off a full definitional chain for f which will interest us the most.

Now, a canonical wff which reflects a full definition chain for f is built up starting from wffs expressing initial wffs. Those starter wffs are Δ_0 wffs – see the proof for E1 – and hence Σ_1.

Suppose g and h are one-place functions, expressed by the Σ_1 wffs $G(x, y)$ and $H(x, y)$ respectively. The function $f(x) = h(g(x))$ is expressed by the wff $\exists z(G(x, z) \land H(z, y))$ – as in the proof for E2 – and this wff is Σ_1 too. Similarly for other cases of composition.

Finally, suppose f is defined from g and h by primitive recursion, where g and h are p.r. functions which can be expressed by the Σ_1 wffs G and H. Then f can be expressed by a wff of the type (3) as constructed in our proof of E3. But since B is Δ_0, it immediately follows that this wff is Σ_1.[4]

So our recipe for building a canonical wff stage by stage takes us from Σ_1 wffs to Σ_1 wffs. Which yields the stronger

Theorem 32 *L_A can canonically express any p.r. function f by a Σ_1 wff which recapitulates a full definitional chain for f.*

10.5 Q can capture all p.r. functions

(a) We have outlined a proof that the language of the theory Q can *express* all p.r. functions using Σ_1 wffs. We now want to show that this theory (and hence any stronger one) can also *capture* all those functions using Σ_1 wffs.

But hold on! We haven't yet officially said what it is for a theory to capture a function. So recall our earlier Defn 19 which tells us what it is to capture a relation, and think of $f(m) = n$ as stating a two-place relation between m and n. That suggests the following definition:

Defn. 34 *The theory T captures the one-place function f by the open wff $\psi(x, y)$ iff, for any m, n,*
  i. if $f(m) = n$, then $T \vdash \psi(\overline{m}, \overline{n})$,
  ii. if $f(m) \neq n$, then $T \vdash \neg\psi(\overline{m}, \overline{n})$.

But (for technical reasons we are not going to fuss about here) it turns out to be useful to replace (ii) with the stronger

  ii′. $T \vdash \exists! y\psi(\overline{m}, y)$.

Here '$\exists!$' is the uniqueness quantifier, 'there exists exactly one'. In other words, T also 'knows' that ψ is functional (associates a number to just one value).

---

[4]Our definition of Σ_1 wffs in §8.3(b) was more complicated than our first-shot sketch. That complication now pays off in making it a matter of observation that this (3)-type wff is Σ_1.

It is elementary that (i) and (ii′) entail (ii). For suppose $f(m) \neq n$. Then, for some k, $f(m) = k$ where $k \neq n$. So by (i) $T \vdash \psi(\overline{m}, \overline{k})$, and also of course $T \vdash \overline{k} \neq \overline{n}$. Together with (ii′) $T \vdash \exists! y \psi(\overline{m}, y)$ simple logic then gives $T \vdash \neg \psi(\overline{m}, \overline{n})$.

(b)   Our target, then, is the following theorem (and compare our earlier announced Theorem 19):

Theorem 33 *The theory* Q *can capture any p.r. function by a* Σ_1 *wff.*

Proof outline   The conceptually easiest route is to use again the same overall strategy we used in proving that Q's language can express every p.r. function. Suppose then that we can prove

C1. Q can capture the initial functions.

C2. If Q can capture the functions g and h, then it can also capture a function f defined by composition from g and h.

C3. If Q can capture the functions g and h, then it can also capture a function f defined by primitive recursion from g and h.

Then it follows that Q can capture any p.r. function.

How do we prove C1? We just check that the formulas which we said in §10.3 *express* the initial functions also serve to *capture* them in Q. That's easy.

How do we prove C2? Suppose g and h are one-place functions, captured by the wffs $G(x, y)$ and $H(x, y)$ respectively. Then we need to prove that the function $f(x) = h(g(x))$ is captured by the wff $\exists z (G(x, z) \wedge H(z, y))$.
   But suppose $f(m) = n$, i.e. for some k, $g(m) = k$ and $k(k) = n$. Then by assumption $T \vdash G(\overline{m}, \overline{k})$ and $T \vdash H(\overline{k}, \overline{n})$, therefore $T \vdash \exists z (G(\overline{m}, z) \wedge H(z, \overline{n}))$. So that gives us clause (i) of the definition of capturing. The uniqueness clause (ii*) follows from the uniqueness clauses for G and H.

How do we prove C3? This is the tedious case that takes hard work! We need to show that B not only expresses but captures Gödel's β-function.[5] And then we use that fact to prove that if the n-place function g is captured by a $(n+1)$-variable expression G, and the $(n+2)$-variable function h by the $(n+3)$-variable expression H, then a wff built to the pattern of (3) in §10.3 captures the function f defined by primitive recursion from g and h. We'll skip over the details, which get messy.[6]
   So take a definition chain for defining a p.r. function. Follow the step-by-step instructions implicit in §10.3 about how to build up a wff which in effect recapitulates that recipe. You'll get a wff that not only canonically expresses but captures the function in Q (and so captures it too in any stronger theory which contains the language of basic arithmetic). And we have already seen that the canonical wff in question will be Σ_1.

---

[5]See again the very end of §10.2. We now see the dodge of using B rather than $\hat{\mathsf{B}}$ coming into its own; the added clause in B is there to reflect the functional character of the β-function.

[6]Not difficult, but messy – as you can see from *IGT2*, particularly §17.3.

10.6 Expressing/capturing properties and relations

We have skipped over quite a few details, but enthusiasts can follow them up elsewhere. All you really need to take away from this chapter so far is a clear grasp of what the announced theorems claim, and enough understanding of the ideas involved in their proofs to grasp too the idea of a p.r. function being expressed/captured by a wff 'canonically', i.e. in a way which recapitulates a definition chain for the function. Then, finally, we need just a brief coda, linking what we've done in this chapter with the last section of the previous chapter.

We said in Defn. 31 that the characteristic function c_P of a monadic numerical property P is defined by setting $c_P(m) = 1$ if m is P and $c_P(m) = 0$ otherwise. And a property P is said to be p.r. decidable if its characteristic function is p.r.

Now, suppose that P is p.r.; then c_P is a p.r. function. So L_A can express c_P by a two-place Σ_1 wff $\mathsf{c}_P(\mathsf{x},\mathsf{y})$. So if m is P, i.e. $c_P(m) = 1$, then $\mathsf{c}_P(\overline{\mathsf{m}},\overline{1})$ is true. And if m is not P, i.e. $c_P(m) \neq 1$, then $\mathsf{c}_P(\overline{\mathsf{m}},\overline{1})$ is not true. Hence, by the definition of expressing-a-property, the wff $\mathsf{c}_P(\mathsf{x},\overline{1})$ serves to express the p.r. property P. The point generalizes from monadic properties to many-place relations. So as an easy corollary of Theorem 32 we get:

Theorem 34 L_A *can express all p.r. decidable properties and relations, again using Σ_1 wffs.*

Similarly, suppose again that the monadic property P is p.r. so c_P is a p.r. function. Then Q can capture c_P by a two-place Σ_1 wff $\mathsf{c}_P(\mathsf{x},\mathsf{y})$. So if m is P, i.e. $c_P(m) = 1$, then $\mathsf{Q} \vdash \mathsf{c}_P(\overline{\mathsf{m}},\overline{1})$. And if m is not P, i.e. $c_P(m) \neq 1$, then $\mathsf{Q} \vdash \neg\mathsf{c}_P(\overline{\mathsf{m}},\overline{1})$. Hence, by the definition of capturing-a-property, the wff $\mathsf{c}_P(\mathsf{x},\overline{1})$ also serves to capture the p.r. property P in Q. The point trivially generalizes from monadic properties to many-place relations. So as an easy corollary of Theorem 33 we get:

Theorem 35 Q *can capture all p.r. decidable properties and relations, again using Σ_1 wffs.*

11 The arithmetization of syntax

Back in Chapter 4, we outlined how Gödel proved his First Incompleteness Theorem. One idea we introduced then was the arithmetization of syntax. We return now to investigate this pivotal idea.

11.1 Gödel-numbering

(a) It was Hilbert[1] who first emphasized that the syntactic objects that comprise formal theories (the wffs, the proofs) are *finite* objects and we only need a mathematics of finite objects to deal with the syntactic properties of theories. We'll return to discuss the significance that this insight had for Hilbert in Chapter 19. But for now, we want Gödel's great twist on this idea: when we are dealing with finite objects, we can give them numerical codes. Hence we can use *arithmetic* to talk – via the coding – about syntactic properties of theories (including the syntactic properties of theories of arithmetic in particular).

Let's start by concentrating on the particular case of coding up expressions of the language L_A. There are various ways of doing this. Here we will follow tradition and use the general style of coding via powers of primes, as used by Gödel himself. But any coding scheme will work for our arguments in the coming chapters, so long as it is 'normal' in a sense which we'll explain shortly.

Suppose, then, that our version of L_A has the usual logical symbols (connectives, quantifiers, identity, brackets), and symbols for zero and for the successor, addition and multiplication functions: associate all those symbols with odd numbers (different symbol, different code number, of course). L_A, as a standard first-order language, also has the usual inexhaustible supply of variables, which we'll associate with even numbers. So, to pin that down, let's fix on this preliminary series of *basic codes*:

$$\neg \quad \wedge \quad \vee \quad \rightarrow \quad \leftrightarrow \quad \forall \quad \exists \quad = \quad (\quad) \quad 0 \quad S \quad + \quad \times \quad x \quad y \quad z \ldots$$
$$1 \quad 3 \quad 5 \quad 7 \quad 9 \quad 11 \quad 13 \quad 15 \quad 17 \quad 19 \quad 21 \quad 23 \quad 25 \quad 27 \quad 2 \quad 4 \quad 6 \ldots$$

Our Gödelian numbering scheme for expressions is now defined in terms of this table of basic codes as follows:

---

[1]David Hilbert (1862–1943) was one of the most influential and wide-ranging mathematicians of the late nineteenth and early twentieth century. Working with his research assistant Paul Bernays, Hilbert's influence on the development of logic was profound.

Defn. 35 *Suppose that the expression e is the sequence $s_1, s_2, \ldots, s_k$ of k symbols and/or variables. Then e's* Gödel number *(g.n.) is calculated by taking the basic code-number c_i for each s_i in turn, using c_i as an exponent for the i-th prime number π_i, and then multiplying the results, to get $2^{c_1} \cdot 3^{c_2} \cdot 5^{c_3} \cdot \ldots \cdot \pi_k^{c_k}$.*

For example:

    i. The single symbol 'S' has the g.n. 2^{23} (the first prime raised to the appropriate power as read off from our correlation table of basic codes).

    ii. The standard numeral SS0 has the g.n. $2^{23} \cdot 3^{23} \cdot 5^{21}$ (the product of the first three primes raised to the appropriate powers).

    iii. The wff $\exists y\,(S0 + y) = SS0$ has the g.n.

$$2^{13} \cdot 3^4 \cdot 5^{17} \cdot 7^{23} \cdot 11^{21} \cdot 13^{25} \cdot 17^4 \cdot 19^{19} \cdot 23^{15} \cdot 29^{23} \cdot 31^{23} \cdot 37^{21}.$$

That last number is, of course, *enormous*. So when we say that it is elementary to decode the resulting g.n. by taking the exponents of prime factors, we don't mean that the computation is quick. We mean that the computational routine required for the task – namely, repeatedly extracting prime factors – involves no more than the mechanical operations of elementary arithmetic. And of course, because numbers are uniquely decomposable into prime factors, the decoding of any number will be unique – either a particular expression or a null result.[2]

(b)   Now, as well as talking about individual wffs via their code numbers, we will also want to talk about whole proofs via *their* code numbers. But how *do* we code for proof-arrays?

The details will obviously depend on the kind of proof system we adopt for the theory we are using. Suppose though, for simplicity, we consider theories with a so-called Hilbert-style axiomatic system of logic. In this rather old-fashioned framework, proof-arrays are simply *linear sequences* of wffs (rather than being, e.g., tree structures). A nice way of coding these linear sequences is by what we'll call *super Gödel numbers*.

Defn. 36 *Given a sequence of wffs or other expressions $e_1, e_2, \ldots, e_n$, we first code each e_i by a regular g.n. g_i, to yield a corresponding sequence of numbers $g_1, g_2, \ldots, g_n$. We then encode this sequence of regular Gödel numbers using a single* super *g.n. by repeating the trick of multiplying powers of primes to get $2^{g_1} \cdot 3^{g_2} \cdot 5^{g_3} \cdot \ldots \cdot \pi_n^{g_n}$.*

Decoding a super g.n. therefore involves two steps of taking prime factors: first find the sequence of exponents of the prime factors of the super g.n.; then treat each of those exponents in turn as a regular g.n., and take prime factors again to arrive back at a sequence of expressions.

---

[2]In the last chapter, we were thinking about how we might handle codes for finite sequences 'inside L_A' – i.e. using the resources of that formal language. In *that* context we couldn't use the device of coding using powers of primes, since L_A lacks a built-in exponential function; so we had to resort to Gödel's β-function trick. But *here* in this chapter we can bring to bear whatever mathematics we like in coding up L_A expressions 'from the outside'.

11.2 The arithmetization of syntactic properties/relations

In this section, we will continue to focus on the language L_A and on the particular theory PA which is built in that language. But as we will stress in the next section, similar results will apply mutatis mutandis to any sensibly axiomatized formal theory.

Recall Defn. 15 from §4.4. In the present context, this becomes:

Defn. 37 *Given our scheme for Gödel-numbering PA expressions and sequences of expressions, we can define the following numerical properties/relations:*

> *Wff*(n) *iff n is the Gödel number of a L_A wff.*
> *Sent*(n) *iff n is the Gödel number of a L_A sentence.*
> *Prf*(m, n) *iff m is the super Gödel number of a PA-proof of the L_A sentence with code number n.*

Then we have the following key result:

Theorem 36 *Wff and Sent are p.r. decidable properties; and Prf is a p.r. decidable relation.*

How do we show this? Writing at the very beginning of the period when concepts of computation were being forged, Gödel couldn't expect his audience to take anything on trust about what was or wasn't '*rekursiv*' or – as we would now put it – primitive recursive. He therefore had to do all the hard work of explicitly showing how to define such properties and relations (or their characteristic functions) by a long chain of definitions by composition and recursion.

However, assuming only a modest familiarity with the ideas of computer programs and p.r. functions, and accepting Theorem 28, we can perhaps short-cut all that effort and be persuaded by the following:[3]

Informal proof sketch Stage one: to determine whether *Wff*(n), first decode n. A simple algorithm does the job, one that doesn't require an open-ended search.

Stage two: ask whether the resulting expression is a wff of the language L_A. That's algorithmically decidable – and again no open-ended search is required. The size of the looping computations required for dealing with the expression will be fixed by its length (which can be computed at stage one).

So neither stage of the decision procedure will involve any open-ended search; both stages can be done by programs using just bounded-in-advance basic 'for' loops.

Now, the second stage of the described informal decision procedure works on strings of symbols. But we can imagine a parallel procedure, operating directly on numbers which code for symbol-strings. So we can think of the whole computation as done on numbers, still without open-ended searches (its only looping structures are more bounded-in-advance basic 'for' loops); so *Wff* is p.r. decidable. Similarly for *Sent*.

---

[3]Frankly, we *are* cutting a few corners here. If you aren't sufficiently persuaded, then be assured that we *can* prove Theorem 36 the hard way, as explained in *IGT2*, Chapter 20.

To determine whether $Prf(m, n)$, first doubly decode the super Gödel number m: that's a mechanical exercise. Now ask: is the result a sequence of PA wffs? That's algorithmically decidable (since it is decidable whether each separate string of symbols is a wff). If it does decode into a sequence of wffs, ask next: is this sequence a properly constructed PA proof? That's decidable too (check whether each wff in the sequence is either an axiom or is an immediate consequence of previous wffs by one of the rules of inference of PA's Hilbert-style logical system). If the sequence is a proof, ask: does its final wff have the g.n. n? That's again decidable. Finally, assuming theorems have to be sentences, ask whether $Sent(n)$ is true.

Putting things together, there is a computational procedure for telling whether $Prf(m, n)$ holds. Moreover, at each stage, the computation involved is once more a straightforward, bounded procedure without any open-ended searches. The length of the sequence of expressions with code m puts a ceiling on the work we have to do and how many times we have to iterate various operations. So the procedure is one that can be written up as a program using only bounded 'for'-loops (and we could in principle do all the computations on numbers rather than symbols). Hence Prf is also a p.r. decidable relation.   ⊠

Which was all rather quick-and-dirty. But surely enough to convince us that Theorem 36 holds good.

There's an immediately corollary. We learnt in the last chapter how PA can express and capture all p.r. properties and relations. Combine those results with our last theorem and we have:

Theorem 37 PA *can express and capture* Wff, *Sent and* Prf *by* Σ_1 *wffs.*

11.3 Generalizing

(a) Theorem 36 tells us that the key relation Prf defined using one particular Gödel-numbering scheme is primitive recursive.

However, our adopted numbering scheme was fairly arbitrarily chosen. We could, for example, shuffle around the preliminary assignment of basic codes to get a different Gödel-style numbering scheme; or we could use a scheme that isn't based on powers of primes. So could it be that a relation like Prf is p.r. when defined in terms of our particular numbering scheme and not p.r. when defined in terms of some alternative but equally acceptable scheme?

Well, what counts as 'acceptable' here? The key feature of our Gödelian scheme is this: there is a pair of algorithms, one of which takes us from an L_A expression to its code number, the other of which takes us back again from the code number to the original expression. Moreover, in following through these algorithms, the upper length of the computation will be determined by the length of the L_A expression to be encoded or by the size of the number to be decoded: *we don't have to go on unbounded searches.* The computations can be done just using bounded 'for' loops.

Generalizing this idea, we will say:

Defn. 38 *A normal Gödel-numbering scheme is one which deploys coding and decoding algorithms which don't involve any open-ended searches.*

Back in Defn. 14, we required the coding and decoding procedures in a Gödel-numbering scheme to be *effective* procedures. We are now tightening this requirement. From now on, we assume that the required effective procedures can be done without open-ended searches, so that our coding schemes are normal. And with this assumption in place, we can simply remark that, whichever (normal) coding scheme for PA we choose, the informal proof we gave will go through as before: *Prf*, now redefined using the new scheme, will still be p.r. decidable.

(b) Now we generalize in a different direction. So: what about Prf_T for other theories T? Take the relation that holds between m and n when (according to our chosen normal numbering scheme) m codes for a T-proof of the sentence coded by n; is this a p.r. relation again for other theories T? ·

Suppose T is not just effectively axiomatized, but – like PA – is put together so that we can mechanically check whether a purported T-proof really is a proof *without* having to set out on an unbounded search. In other words, suppose that checking a proof can be done by a procedure that can be regimented using nothing more exotic than 'for' loops. Then, by the same informal proof as before, the relation Prf_T which holds when m numbers a proof of the wff with number n will still be primitive recursive.

Let's say that

Defn. 39 *A theory T is* p.r. axiomatized *when it is so axiomatized as to make the relation Prf_T (defined using a normal Gödel-numbering scheme) a primitive recursive relation.*

Then *any* usual kind of formal theory you dream up will actually be p.r. axiomatized if it is effectively axiomatized at all. We *never* in practice formalize a theory in such a way that (i) it is effectively axiomatized, but (ii) we can only effectively check whether a given array of expressions is a well-constructed proof by some unbounded search(es). Hold onto that important observation!

11.4 Some cute notation

We now introduce a rather pretty bit of notation. Assume we have chosen some system for Gödel-numbering the expressions of a language L. Then

Defn. 40 *If φ is an L-expression, then we'll use '$\ulcorner \varphi \urcorner$' in our logicians' augmented English to denote φ's Gödel number. And we use '$\overline{\ulcorner \varphi \urcorner}$' as an abbreviation in our formal arithmetic for the standard numeral for the number $\ulcorner \varphi \urcorner$.*

Borrowing corner quotes for this new use is quite appropriate because the number $\ulcorner \varphi \urcorner$ can be thought of as referring to the expression φ via our coding scheme. (Sometimes, we'll write the likes of $\ulcorner U \urcorner$ where U abbreviates an L_A wff: we then mean, of course, the Gödel-number for the original wff that U stands in for.)

And given we used the overlined expression '$\overline{n}$' to abbreviate the standard numeral for the number n, it is quite natural to use the same convention again in using the overlined '$\overline{\ulcorner\varphi\urcorner}$' to abbreviate the standard numeral for $\ulcorner\varphi\urcorner$. So:

(1) 'SS0' is an L_A expression, the standard numeral for 2.

(2) On our numbering scheme $\ulcorner$SS0$\urcorner$, the g.n. of 'SS0', is $2^{23} \cdot 3^{23} \cdot 5^{21}$.

(3) Then '$\overline{\ulcorner$SS0$\urcorner}$' is shorthand for the standard L_A-numeral for that g.n., i.e. as an abbreviation for 'SSS...S0' with $2^{23} \cdot 3^{23} \cdot 5^{21}$ occurrences of 'S'.

11.5   Diagonalization

Finally, we use our new notation in defining a simple construction which will play an all-important role in the coming chapters (so read carefully!).

Assume we are working with a language with standard numerals, and that we have a Gödel-numbering scheme in play: then

Defn. 41 *The diagonalization of a wff φ with one free variable is the wff $\varphi(\ulcorner\varphi\urcorner)$*

Let's clarify: now making the free variable explicit, the diagonalization of the wff $\varphi(x)$ is what you get by replacing occurrences of its free variable x by the numeral for the Gödel number of the *whole* wff $\varphi(x)$. For example, the diagonalization of $\exists y F(x, y)$ is $\exists y F(\ulcorner \exists y F(x, y) \urcorner, y)$.

Why is this substitution operation called *diagonalization*? Well compare the 'diagonal' construction we encountered in §5.4. There, we counted off wffs $\varphi_0(x)$, $\varphi_1(x)$, $\varphi_2(x)$ ... in an enumeration of wffs with x as their one free variable; and then we substituted (the numeral for) the index d for the free variable in the wff φ_d, to form $\varphi_d(\overline{d})$. We can now think of the Gödel number of a wff with one free variable as another way of indexing that wff in a list of such wffs. And so, in our new diagonal construction, we are again substituting (the numeral for) the index of a wff for the free variable in the wff.

An important observation. Diagonalization is an elementary mechanical operation on expressions. Hence, assuming that our Gödel numbering scheme for the language L is normal, we can expect this to be true:

Theorem 38 *There is a p.r. function $diag(n)$ which, when applied to a number n which is the g.n. of some L-wff with one free variable, yields the g.n. of that wff's diagonalization, and yields 0 otherwise.*

As with any p.r. function, T can express and capture this function by a Σ_1 wff, which we will abbreviate $\mathsf{Diag}_T(x, y)$.

Proof Try treating n as a g.n., and seek to decode it. If you don't get an expression with one free variable, return 0. Otherwise you get a wff of the type $\varphi(x)$ (the particular free variable doesn't matter), and can then form the wff $\varphi(\overline{n})$, which is its diagonalization (since by assumption n is its g.n.). Then we work out the g.n. of this resulting wff to compute $diag(n)$.

This procedure doesn't involve any unbounded searches. So we will be able to program the procedure using just 'for' loops. Hence *diag* is a p.r. function.

Now apply the key theorems of Chapter 10.   ☒

12 The First Incompleteness Theorem, semantic version

Let's quickly review some crucial ideas that we have met in the last couple of chapters:

i. We fixed on a particular scheme for coding up wffs of PA's language L_A by using Gödel numbers ('g.n.' for short), and for coding up PA-proofs by super Gödel numbers (assuming for convenience that these proofs are simple sequences of wffs). On our scheme, the algorithms which take us from expressions to code numbers and back again don't involve any open-ended searches. Call a coding scheme with this feature *normal*. (§11.2)

ii. Given our coding scheme, we can define properties and relations such as $Prf(m, n)$, which is the relation that holds if and only if m is the super g.n. of a sequence of wffs that is a PA proof of a sentence with g.n. n. Such properties and relations are primitive recursive. (§11.2)

iii. Any p.r. function, property or relation can be *expressed* by a wff of PA's language L_A. For example, we can construct a Σ_1 wff Prf which *canonically* expresses *Prf* by recapitulating a definition chain for the relation's characteristic function. (§§4.5, 10.3)

iv. Any p.r. function, property or relation can be *captured* in Q and hence in PA. For example, *Prf* can be captured by a Σ_1 wff (again one which recapitulates the relation's p.r. definition). (§10.5)

v. Notation: If φ is an expression, then we'll denote its Gödel number in our logician's English by '$\ulcorner\varphi\urcorner$'. We use '$\overline{\ulcorner\varphi\urcorner}$' as an abbreviation for the standard numeral for $\ulcorner\varphi\urcorner$. (§11.4)

vi. The diagonalization of a wff φ with one free variable is $\varphi(\overline{\ulcorner\varphi\urcorner})$. The function *diag* sends the g.n. of a wff to the g.n. of its diagonalization and is primitive recursive. (§11.5)

For what follows, it isn't necessary that you know the *proofs* of the claims we've just summarized: but do pause to double check that you at least fully understand what the various claims *mean*. And then, read on ...

12.1 Constructing a Gödel sentence

(a) As we announced in Chapter 4, Gödel is going to tell us how to construct a wff G that is true if and only if it is unprovable in PA. We now have an inkling of how he can do that. Arithmetization allows us to construct wffs which express proof-relations like *Prf*. Moreover, wffs can contain numerals which refer to numbers which – via Gödel coding – are in turn correlated with wffs. Maybe, if we are cunning, we can get a wff to be 'about' itself and its own unprovability.

And at the end of the last chapter we met the basic construction we are going to use, namely diagonalization. However, simply inserting the numeral for an open wff W into that same wff gives us a new wff W^* which is not 'about' *itself* but 'about' W. We will have to be a bit cleverer.

We start, then, by tweaking the definition of *Prf* to get

Defn. 42 *The relation $Prfd(m, n)$ holds if and only if m is the super g.n. for a* PA *proof of the diagonalization of the wff with g.n. n.*

Suppose n Gödel-numbers a wff $\varphi(x)$; then $Prfd(m, n)$ holds if m super-Gödel-numbers a proof of the wff $\varphi(\bar{n})$, which is the diagonalization of φ. Equivalently, $Prfd(m, n)$ holds if and only if m stands in the *Prf* relation to $diag(n)$.

Theorem 39 *Prfd is primitive recursive.*

Proof (Informal) We just remark that, as with $Prf(m, n)$, we can mechanically check whether $Prfd(m, n)$. Just decode m. Check whether it gives a sequence of wffs. If it does, check whether it is a PA proof. If it is, check whether the concluding wff of the proof is the diagonalization of the wff with g.n. n. That involves no open-ended searches. An algorithm involving only basic 'for' loops will suffice. So *Prfd* is primitive recursive. ⊠

Proof (Official) $Prfd(m, n)$ is true iff $Prf(m, diag(n))$. Let c_{Prf} be the characteristic function of *Prf*, which is primitive recursive. Then the characteristic function of *Prfd* is $c_{Prfd}(x, y) =_{def} c_{Prf}(x, diag(y))$, so – being a composition of p.r. functions – c_{Prfd} is primitive recursive too. ⊠

Since *Prfd* is p.r., it can be expressed and captured in PA by some Σ_1 wff. And for present purposes, we don't really need to say more about the construction of the relevant wff: it is enough for there to be a wff which does the job.

But we can in fact be more specific. For let the relation *Prf* be canonically expressed and captured by the Σ_1 wff $\mathsf{Prf(x, y)}$; and let the function *diag* be likewise expressed and captured by the Σ_1 wff $\mathsf{Diag(x, y)}$. Then, since $Prfd(m, n)$ is simply $Prf(m, diag(n))$, this is easily checked:[1]

Theorem 40 *Prfd can be canonically expressed and captured by the Σ_1 wff* $\mathsf{Prfd(x, y)} =_{def} \exists \mathsf{z(Prf(x, z) \land Diag(y, z))}$.

And we'll fix on that as our official definition.

---

[1]Compare the proofs for (E2) in §10.3 and (C2) in §10.5.

(b) Now comes the ingenious Gödelian construction! First we form the open wff we'll abbreviate as U (think 'unprovable'). To make its free variable explicit,

Defn. 43 $U(y) =_{def} \neg\exists x\, Prfd(x, y)$.

Now we diagonalize U, to give

Defn. 44 $G =_{def} U(\ulcorner U\urcorner) = \neg\exists x\, Prfd(x, \ulcorner U\urcorner)$.

And here is the wonderful result:

Theorem 41 G *is true if and only if it is unprovable in* PA.

Proof Consider what it takes for G to be true (on the interpretation built into L_A), given that the formal predicate Prfd expresses the numerical relation *Prfd*.

G, i.e. $\neg\exists x\, Prfd(x, \ulcorner U\urcorner)$, is true if and only if no number m is such that $Prfd(m, \ulcorner U\urcorner)$. That is to say, given the definition of *Prfd*, G is true if and only if there is no number m such that m is the code number for a PA proof of the diagonalization of the wff with g.n. $\ulcorner U\urcorner$. But the wff with g.n. $\ulcorner U\urcorner$ is of course U; and its diagonalization is G.

So, G is true if and only if there is no number m such that m is the code number for a PA proof of G. But if G is provable in PA, some number would be the code number of a proof of it. Hence G is true if and only if it is unprovable in PA. ⊠

(c) G – meaning of course the L_A sentence you get when you unpack the abbreviations! – is thus our promised Gödel sentence for PA.

And we might call this a *canonical* Gödel sentence for PA for three reasons: (a) it is defined in terms of a wff that we said canonically expresses/captures *Prfd*, and (b) because it is the sort of sentence that Gödel himself constructed, so (c) it is the kind of sentence people standardly have in mind when they talk of '*the*' Gödel sentence for PA.

It is true that G will be horribly long when spelt out in quite unabbreviated L_A with its uneconomical standard numerals. But in another way, it is relatively simple. For G is of Goldbach type (in the sense of §8.3) – or in other words:

Theorem 42 G *is* Π_1.

Proof $Prfd(x, y)$ is Σ_1. So $Prfd(x, \ulcorner U\urcorner)$ is still Σ_1 (we've just filled up one slot in the open wff with a numeral). So is its existential quantification $\exists x\, Prfd(x, \ulcorner U\urcorner)$ (the result of adding another existential quantifier to the front of a Σ_1 wff is still Σ_1). Negating to get G gives us a Π_1 wff (the negation of a Σ_1 wff is Π_1). ⊠

12.2 What G says

It is often claimed that a canonical Gödel sentence like G actually *says* of itself that it is unprovable. However, that can't be strictly true.

G (when unpacked) is just another sentence of PA's language L_A, the language of basic arithmetic. It is a long wff involving the first-order quantifiers, the connectives, the identity symbol, and 'S', '+' and '×', which all have the standard interpretation built into L_A. In particular, the standard numeral in G refers to a number, not a wff; and the quantifier in G runs over numbers. So G strictly speaking says something about *numbers*, not about wffs and their unprovability.

However there is perhaps a fairly reasonable sense in which G *can* be described as *indirectly* saying that it is unprovable. Note, this is *not* to make play with some radical re-interpretation of G's symbols (that would just give us a boring triviality: if we are allowed radical re-interpretations – like spies choosing to borrow ordinary words for use in a secret code – then any string of symbols can be made to say anything). No, it is because the symbols are still being given their *standard* interpretation that we can recognize that the canonically constructed Prfd (when unpacked) will express *Prfd*, given the background framework of Gödel numbering which is involved in the definition of the relation *Prfd*. Therefore, given that coding scheme, we can recognize just from its construction that G will be true when no number m is such that $Prfd(m, \ulcorner U \urcorner)$, and so no number codes for a proof of G. In short, given the coding scheme, we can see just from the way it is constructed that G is true just when it is unprovable. *That* is the limited sense in which, via our Gödel coding, the canonical G signifies or 'indirectly says' that it is unprovable.

12.3 The First Theorem for PA – the semantic version

We already announced in §4.6 that Gödel tells us how to construct a wff which is true if and only if it is unprovable. And as we showed then, the argument to an incompleteness theorem is now very straightforward. Here it is again.

Assume PA is sound, i.e. proves no falsehoods (because its axioms are true and its logic is truth-preserving). If G could be proved in PA, then PA *would* prove a false theorem (since G is true iff it is *not* provable). That would contradict our supposition that PA is sound. Hence, G is not provable in PA.

But that shows that G *is* true. So ¬G must be false. Hence ¬G cannot be proved in PA either, assuming PA is sound. In Gödel's words, then, G is a 'formally undecidable' sentence of PA (see Defn. 7):

Theorem 43 *If* PA *is sound, then there is a true* Π_1 *sentence* G *such that* PA $\nvdash$ G *and* PA $\nvdash$ ¬G, *so* PA *is negation incomplete.*

If we are happy with the semantic assumption that PA's axioms *are* true on interpretation and so PA *is* sound, the argument for incompleteness is as simple as that – or at least, it's that simple once we have constructed G.

12.4 Generalizing the proof

This line of proof now generalizes. Suppose T is any theory which (i) contains the language of basic arithmetic (see Defn. 11), so T can form standard numerals, and we can form the diagonalization of a T-wff with one free variable. Suppose (ii) that we are working with a normal system of Gödel-numbering for T-wffs and

T-proofs. Suppose also (iii) that T is p.r. axiomatized in the sense of Defn. 39. Then we can again define a relation $Prfd_T(m, n)$ which holds when m numbers (on our new scheme) a T-proof of the diagonalization of the wff with number n; and this relation will be primitive recursive.

Continuing to suppose that T's language includes the language of basic arithmetic, T will be able to express the p.r. relation $Prfd_T$ by a Σ_1 wff Prfd_T. Then, just as we did for PA, we'll be able to construct the corresponding Π_1 wff G_T which is true if and only if it is not provable in T. Then, by exactly the same argument as before we can show, more generally,

Theorem 44 *If T is a sound p.r. axiomatized theory whose language contains the language of basic arithmetic, then there will be a true Π_1 sentence G_T such that $T \nvdash \mathsf{G}_T$ and $T \nvdash \neg\mathsf{G}_T$, so T is negation incomplete.*

So this is our first, '*semantic*', version of the generalized First Incompleteness Theorem!. And let's note an immediate corollary:

Theorem 45 *There is no p.r. axiomatized theory framed in the language of L_A whose theorems are all and only the truths of L_A.*

For if the theorems are all true, our theory is sound, and it can't be complete.

12.5 Our Incompleteness Theorem is better called an *incompletability* theorem

Here, we just revisit the argument of §3.3: but the point is crucial enough to bear repetition. Suppose T is a sound p.r. axiomatized theory which can express claims of basic arithmetic. Then by Theorem 44 we can find a true G_T such that $T \nvdash \mathsf{G}_T$ and $T \nvdash \neg\mathsf{G}_T$. That *doesn't* mean that G_T is 'absolutely unprovable' (it is *very* unclear whether we can give any good sense to that notion): it just means that G_T is unprovable-in-T.

OK: let's simply augment T by adding G_T as a new axiom, to give the theory $U = T + \mathsf{G}_T$. Then (i) U is still sound (for the old T-axioms are true, the added new axiom is true, and the logic is still truth-preserving). (ii) U is evidently still a p.r. axiomatized theory (why?). (iii) We haven't changed the language. So our Incompleteness Theorem applies, and we can find a sentence G_U such that $U \nvdash \mathsf{G}_U$ and $U \nvdash \neg\mathsf{G}_U$.[2] Since U can prove everything T can prove, that implies $T \nvdash \mathsf{G}_U$ and $T \nvdash \neg\mathsf{G}_U$. In other words, as we put it before, 'repairing the gap' in T by adding G_T as a new axiom leaves some other sentences that are undecidable in T *still* undecidable in the augmented theory.

In sum, our incompleteness theorem tells us that if we keep chucking more and more additional axioms at T, our theory will still remain negation-incomplete, unless it either stops being sound or stops being p.r. axiomatized. In a good sense, T is *incompletable*.

---

[2]Note, U is a different theory to T, hence $Prfd_U$ is a different relation to $Prfd_T$, and so G_U (although constructed in a parallel way) will indeed be a different wff to G_T.

12.6   Comparing old and new semantic incompleteness theorems

Finally in this chapter, let's compare the new Theorem 44 which we *have* just proved with the old theorem which we initially announced in §3.2 but of course *didn't* there prove:

Theorem 1 *Suppose T is an effectively axiomatized formal theory whose language contains the language of basic arithmetic. Then, if T is sound, there will be a true sentence G_T of basic arithmetic such that $T \nvdash \mathsf{G}_T$ and $T \nvdash \neg\mathsf{G}_T$, so $T is negation incomplete.*

Our new theorem is stronger than the announced old one in one respect, weaker in another. But the gain is much more than the loss.

Our new theorem is stronger, because it tells us much more about the character of the undecidable Gödel sentence – namely it has fairly minimal quantifier complexity. The unprovable sentence G_T is a Π_1 sentence of arithmetic. We'll return to say more about this in the next chapter.

Our new theorem is weaker, however, as it only applies to p.r. axiomatized theories, not to (effectively) axiomatized theories more generally. And as we will see in §13.6, Gödel's own original theorems strictly speaking only apply to p.r. axiomatized theories. But that's no real loss. After all, what would a theory look like that was effectively axiomatized but *not* p.r. axiomatized? It would mean that we could e.g. only tell what's an axiom on the basis of an open-ended search: but that would require an *entirely* unnatural way of specifying the theory's axioms in the first place. As we noted at the end of §11.3, any normally presented effectively axiomatized theory will be p.r. axiomatized.

So while we *can* go on to beef up our new result to make it apply as generally as the announced Theorem 1 – and we will say more about this in §17.5 – there is relatively limited interest in doing so. The real force of the (semantic) First Incompleteness Theorem is already captured by Theorem 44.

13 The First Incompleteness Theorem, syntactic version

We now use the same construction of a Gödel sentence as in the previous chapter to show again that PA is incomplete, but this time we proceed making only syntactic assumptions. And then we show how to generalize this syntactic version of the incompleteness theorem.

13.1 ω-completeness, ω-consistency

(a)   We need to define two key notions. We'll assume in this section that we are dealing with theories whose language includes the language of basic arithmetic. And take the quantifiers mentioned to run over the natural numbers.[1]

First, then,

Defn. 45 *A theory T is ω-incomplete iff, for some open wff $\varphi(x)$, T can prove $\varphi(\overline{n})$ for each natural number n, but T can't go on to prove $\forall x\varphi(x)$.*

We saw in §6.8 that Q is ω-incomplete: that's because it can prove each instance of $0 + \overline{n} = \overline{n}$, but can't prove $\forall x(0 + x = x)$. We added induction to Q hoping to repair as much ω-incompleteness as we could: but, as we'll see, PA remains ω-incomplete, assuming it is consistent.[2]

Second, we want the following idea:

Defn. 46 *A theory T is ω-inconsistent iff, for some open wff $\varphi(x)$, T can prove $\varphi(\overline{n})$ for each n and T can also prove $\neg\forall x\varphi(x)$. Equivalently, T is ω-inconsistent iff, for some open wff $\psi(x)$, T can prove each $\neg\psi(\overline{n})$ and T can also prove $\exists x\psi(x)$.*

Note that ω-inconsistency, like ordinary inconsistency, is a syntactically defined property: it is characterized in terms of what wffs can be proved by the theory, not in terms of what the wffs mean.

Note too that, at least is standard logic, ω-consistency – defined of course as not being ω-inconsistent! – trivially implies plain consistency. That's because

---

[1] If necessary, read $\forall x\varphi(x)$ as abbreviating a restricted quantifier $\forall x(Nx \to \varphi(x))$, where 'N' picks out the numbers from the domain of the theory's native quantifiers (see Defn. 11).

[2] Why the 'ω' in 'ω-incomplete'? Because 'ω' is a standard label for the set of natural numbers (when equipped with their usual ordering).

T's being ω-consistent is a matter of its *not* being able to prove a certain combination of wffs, which entails that T can't prove *all* wffs, and hence T can't be inconsistent.

(b) Now compare and contrast. Suppose T can prove $\varphi(\overline{n})$ for each n. T is ω-incomplete if it can't prove something we'd then also like it to prove, namely $\forall x \varphi(x)$. While T is ω-inconsistent if it can actually prove the *negation* of what we'd like it to prove, i.e. it can prove $\neg \forall x \varphi(x)$.

So ω-incompleteness in a theory of arithmetic is a regrettable weakness. But ω-inconsistency is a Very Bad Thing (not as bad as outright inconsistency, maybe, but still bad enough). Evidently, a theory T that can prove each of $\varphi(\overline{n})$ and yet also proves $\neg \forall x \varphi(x)$ is just not going to be an acceptable candidate for regimenting arithmetic.

Bring semantic ideas back into play for a moment. Suppose T's standard numerals denote the numbers and the quantifier here runs over all and only the natural numbers. Then it can't be the case that each of $\varphi(\overline{n})$ is true and yet $\neg \forall x \varphi(x)$ is true too. So our ω-inconsistent T can't be sound.

Given that we want formal arithmetics to have theorems which *are* all true on a standard interpretation, we must therefore want ω-consistent arithmetics. And given that we think e.g. PA *is* sound on its standard interpretation, we are committed to thinking that it *is* ω-consistent.

13.2   The First Theorem for PA – the syntactic version

G is by definition the diagonalization of the open wff U $=_{\text{def}} \neg \exists x\, \text{Prfd}(x, y)$, i.e. G is the wff $\neg \exists x\, \text{Prfd}(x, \ulcorner U \urcorner)$ (see §12.1). And now recall Theorem 40: the wff Prfd doesn't just express *Prfd* but *captures* it. Using this fact about Prfd, we can again show that PA does not prove G, but this time *without* making the semantic assumption that PA is sound:

Theorem 46 *If* PA *is consistent,* PA $\nvdash$ G.

Proof Suppose that PA $\vdash$ G, i.e. (i) PA $\vdash \neg \exists x\, \text{Prfd}(x, \ulcorner U \urcorner)$.

If G has a proof, then there is some super g.n. m that codes its proof. But G is the diagonalization of the wff with g.n. $\ulcorner U \urcorner$. Hence, by definition, $Prfd(m, \ulcorner U \urcorner)$. Since Prfd captures the relation *Prfd*, it follows that PA $\vdash \text{Prfd}(\overline{m}, \ulcorner U \urcorner)$. So, existentially quantifying, we have (ii) PA $\vdash \exists x\, \text{Prfd}(x, \ulcorner U \urcorner)$.

Hence, combining (i) and (ii), the supposition that PA $\vdash$ G entails that PA is inconsistent. Therefore, if PA is consistent, PA $\nvdash$ G.   ⊠

We'll now show that PA also can't prove the *negation* of G, again without assuming PA's soundness:

Theorem 47 *If* PA *is ω-consistent,* PA $\nvdash \neg$G.

Proof Suppose PA is ω-consistent and PA $\vdash \neg$G. We derive a contradiction, and the theorem follows.

Given our supposition that PA is ω-consistent, it is consistent. Hence, given our supposition that PA proves ¬G, it can't prove G. So no number m is the super g.n. of a proof for G. But, again, G is the diagonalization of the wff with g.n. $\ulcorner U \urcorner$. Hence, for every number m, $Prfd(m, \ulcorner U \urcorner)$ is false.

Since Prfd captures the relation $Prfd$ and $Prfd(m, \ulcorner U \urcorner)$ is false for each m, by the definition of capturing we have (i) for each m, PA $\vdash$ ¬Prfd($\overline{m}, \overline{\ulcorner U \urcorner}$).

But our supposition PA $\vdash$ ¬G is equivalent to (ii) PA $\vdash$ ∃xPrfd(x, $\ulcorner U \urcorner$).

Combining (i) and (ii), PA is ω-inconsistent, contradicting our initial supposition. $\boxtimes$

Now recall that G is a Π_1 sentence. Putting that observation together with what we've shown in this section gives us the following portmanteau result:

Theorem 48 *If* PA *is consistent, then there is a* Π_1 *sentence* G *such that* PA $\nvdash$ G, *and if* PA *is* ω-*consistent* PA $\nvdash$ ¬G. *Hence, assuming* ω-*consistency and so consistency,* PA *is negation incomplete.*

13.3   Two corollaries

(a)   We pause to note two corollaries. The first is very important:

Theorem 49 *If* PA *is consistent, it is* ω-*incomplete.*

Proof Assume PA is consistent. By Theorem 46, it doesn't prove G. So no number m is the super g.n. of a proof of G. Hence, exactly as in the proof of Theorem 47, (i) for each m, PA $\vdash$ ¬Prfd($\overline{m}, \ulcorner U \urcorner$). But the Theorem that G is unprovable in PA is trivially equivalent to (ii) PA $\nvdash$ ∀x¬Prfd(x, $\ulcorner U \urcorner$).

The combination of (i) and (ii) shows that PA is ω-incomplete. $\boxtimes$

Let's linger over this result. The incompleteness theorem tells us that there are truths of basic arithmetic which PA can't prove. Reporting it like *that*, however, leaves open the possibility that these will be recondite truths to be found in some remote corner of arithmetic of little interest. But not so. Gödel has found incompleteness surprisingly close to home, in what PA can prove about effectively computable properties of numbers. In particular,

> There's a certain effectively decidable property F which in fact every number has. As we'd hope of a competent formal theory of arithmetic, PA can track what we can compute – so it too can correctly prove of any given number m that m is F. But PA can't take what would seem to be the *very* modest extra step of proving the explicit generalization that *every* number is F.

We have an effectively computable relation *Prfd* (which with only a little effort can be provided with a definitional chain showing it to be primitive recursive); and so there is an effectively computable property F of *not* standing in the *Prfd* relation to the particular number $\ulcorner U \urcorner$. We now know that every number has property F. And indeed (i), for each particular m, PA can correctly prove ¬Prfd($\overline{m}, \ulcorner U \urcorner$). But PA can't take the extra step (ii) and prove the corresponding universal quantification ∀x¬Prfd(x, $\ulcorner U \urcorner$). This sort of incompleteness – as we will see – also affects any system of arithmetic satisfying some natural conditions. Which is surely remarkable.[3]

(b)   For future use, here is the other corollary we want to mention:

Theorem 50 *If* PA *is consistent, then* PA $+ \neg$G *(the theory you get by adding* $\neg$G *as an additional axiom) is also consistent but is* ω-*inconsistent.*

Proof Assume PA is consistent. If PA $+ \neg$G were inconsistent, then PA would prove G, contrary to Theorem 46. So PA $+ \neg$G is also consistent.

Now, since the expanded theory can prove everything PA can prove, we know as before that (i) for each m, PA $+ \neg$G $\vdash \neg$Prfd$(\overline{m}, \ulcorner$U$\urcorner)$.

But just by the definition of $\neg$G, (ii) PA $+ \neg$G $\vdash \exists$xPrfd$(x, \ulcorner$U$\urcorner)$.

And (i) and (ii) together imply that PA $+ \neg$G is ω-inconsistent.   ⊠

13.4   Generalizing the proof

(a)   The proof for Theorem 48 evidently generalizes. Suppose T is a p.r. axiomatized theory which (perhaps after introducing some new vocabulary via abbreviatory definitions) *contains* Q – in the obvious sense that the language of T includes the language of basic arithmetic, and T can prove every Q-theorem. Then, assuming we are working with a normal scheme for Gödel-numbering wffs of T, the relation $Prfd_T(m, n)$ which holds when m numbers a T-proof of the diagonalization of the wff with number n will be primitive recursive again.

Since T can prove everything Q proves, T will be able to capture the p.r. relation $Prfd_T$ by a Σ_1 wff Prfd$_T$. Just as we did for PA, we'll be able to construct the corresponding Π_1 wff G$_T$. So exactly the same arguments as before will then show, more generally,

Theorem 51 *If* T *is a consistent p.r. axiomatized theory which contains* Q, *then there will be a* Π_1 *sentence* G$_T$ *such that* $T \nvdash$ G$_T$, *and if* T *is* ω-*consistent,* $T \nvdash \neg$G$_T$. *Hence, assuming* ω-*consistency and so consistency,* T *is negation incomplete.*

T will also be ω-incomplete in the same way as PA is.

And note, this gives us another *incompletability* theorem: if we keep chucking more and more additional axioms at our theory T, it will still remain negation incomplete, unless it stops being ω-consistent or stops being p.r. axiomatized. In other jargon, it is said that T is *essentially incomplete*.

(b)   When people refer to the *First Incompleteness Theorem* (without qualification), they typically mean something like our last Theorem, deriving incompleteness from *syntactic* assumptions. Let's re-emphasize that last point. Being

---

[3]In his classic book on *Gödel's Incompleteness Theorems* (OUP 1992, p. 73), Raymond Smullyan remarks that, while the average mathematician who isn't a logician has heard about theories of arithmetic not being able to decide everything, relatively few have heard of this remarkably basic level of ω-incompleteness.

p.r. axiomatized is a syntactic property; containing Q is a matter of Q's axioms again being adopted as axioms or being derivable, another syntactic property; being consistent here is a matter of no contradictory pair φ, $\neg\varphi$ being derivable; being ω-consistent is another syntactic property as we stressed before. The chains of argument that lead to our Theorem depend just on the given syntactic assumptions, via e.g. the proof that Q can capture all p.r. functions – another claim about a syntactically definable property. That is why we are calling this the *syntactic* incompleteness theorem.

Of course, we are *interested* in these various syntactically definable properties because of their semantic relevance: for example, we care about the idea of capturing p.r. functions because we are interested in what an interpreted theory might be able to prove in the sense of establish-as-true. But it is one thing for us to have a semantic *motivation* for being interested in a certain concept, it is another thing for that concept actually to have semantic content. Again, we are guided to the construction of a Gödel sentence G_T by considerations of what it as-it-were says: but its interpretation isn't involved at all in the proof of its formal undecidability from syntactic assumptions.

13.5 Comparing old and new syntactic incompleteness theorems

Compare Theorem 51 with our initially announced

Theorem 2 *Suppose T is an effectively axiomatized formal theory whose language contains the language of basic arithmetic. Then, if T is consistent and can prove a certain modest amount of arithmetic (and has an additional property that any sensible formalized arithmetic will share), there will be a sentence G_T of basic arithmetic such that $T \nvdash G_T$ and $T \nvdash \neg G_T$, so T is negation incomplete.*

Our new theorem fills out the old one in various respects. It fixes the 'modest amount of arithmetic' that T is assumed to contain and it also spells out the 'additional desirable property' of ω-consistency which we previously left mysterious. Further it tells us more about the undecidable Gödel sentence – namely it has minimal quantifier complexity, i.e. it is a Π_1 sentence of arithmetic. Our new theorem is weaker, however, as it only applies to p.r. axiomatized theories, not to effectively axiomatized theories more generally. But as we've already noted at the end of the last chapter, that's not much loss. (And again, if we insist, we can in fact go on to make up the shortfall, as we will see.)

13.6 Gödel's own Theorem

To repeat, Theorem 51, or something very like it, is what people usually mean when they speak without qualification of 'The First Incompleteness Theorem'. But since the stated theorem refers to Robinson Arithmetic Q (developed by Robinson in 1950), and Gödel didn't originally know about that (in 1931), our version can't be quite what Gödel originally proved. But it is very close. Let's explain.

Looking again at our analysis of the syntactic argument for incompleteness, we see that we are interested in theories which extend Q *because we are interested in theories which can capture p.r. relations like Prfd*. It's being able to capture *Prfd* that is the crucial condition for our proof to go through. So let's say

Defn. 47 *A theory T is* p.r. adequate *if it can capture all primitive recursive functions and relations.*

Then, instead of mentioning Q, let's instead explicitly write in the requirement of p.r. adequacy. So, by just the same arguments,

Theorem 52 *If T is a p.r. adequate, p.r. axiomatized theory whose language includes L_A, then there is a Π_1 sentence φ such that, if T is consistent then $T \nvdash \varphi$, and if T is ω-consistent then $T \nvdash \neg\varphi$.*

And *this* is pretty much Gödel's own general version of the incompleteness result, and has as much historical right as any to be called *Gödel's* First Theorem.[4]

Thus, in his 1931 paper, Gödel proves that a particular formal system P – which is his simplified version of the hierarchical type-theory of *Principia Mathematica* – has a formally undecidable Π_1 sentence (or as he puts it, the undecidable sentence is of Goldbach type). He then immediately generalizes:

> In the proof of [his key result about P] no properties of the system P were used besides the following:
>
> (1) The class of axioms and the rules of inference (that is, the relation 'immediate consequence') are [primitive] recursively definable (as soon as we replace the primitive signs in some way by the natural numbers).
>
> (2) Every [primitive] recursive relation is definable [i.e., in our terms, is 'capturable'] in the system P.
>
> Therefore, in every formal system that satisfies the assumptions 1 and 2 and is ω-consistent, there are undecidable propositions of the form $[\forall x F(x)]$, where F is a [primitive] recursively defined property of natural numbers, and likewise in every extension of such a system by a recursively definable ω-consistent class of axioms.

Which gives us a version of our Theorem 52.

---

[4] "Hold on! If *that's* the original First Theorem, we don't need to do all the extra work showing that Q and PA are p.r. adequate, do we?" Yes and no! No, proving *this* original version of the Theorem of course doesn't depend on proving that any particular theory is p.r. adequate. But yes, showing that this Theorem has real bite, showing that it does actually apply to arithmetics like PA, does depend on proving that these arithmetics are indeed p.r. adequate.

Interlude

Time to take stock again. In the last Interlude, we signposted the route forward to our two principal versions of the First Theorem. So, to recap:

Chapter 9. We defined a large subclass of the effectively computable functions, namely the primitive recursive (p.r.) ones. We also defined the p.r. properties and relations – those whose characteristic functions are primitive recursive.

We didn't explore very far; but it is more or less enough for our purposes to carry away the thought that the p.r. functions are those which can be effectively computed without open-ended searches.

Chapter 10. Next, we showed that any language as rich as the language of Q can *express* all p.r. functions, properties and relations using just Σ_1 wffs. Moreover any theory which includes Q can also *capture* all those functions etc., using the same kind of wffs.

The ideas shaping the proofs of these two key results are really rather neat: we take a definition chain for a p.r. function and build-up a wff for expressing/capturing the wff step-by-step, tracking the definition chain. And to handle the steps where we define a more complex function from simpler ones by primitive recursion, we use the β-function trick. The devil, of course, is in the details!

Chapter 11. We revisited the idea of the 'arithmetization of syntax' by Gödel-numbering. We then defined various numerical properties/relations such as $Prf_T(m, n)$ (which holds when m is the code number of a T-proof of the sentence with code number n). For sensible theories T, we can effectively decide when such properties/relations obtain, and without needing open-ended searches: in other words, the likes of Prf_T are p.r. (so can be expressed/captured by a Σ_1 wff Prf_T).

We also defined the idea of diagonalization (taking a wff with one free variable, and substituting the numeral for its Gödel number for the variable). The function $diag_T(n)$ which returns the code-number of the diagonalization of the wff with code number n if that exists, and 0 otherwise, is also p.r.

There was a fair amount of ground covered in these three chapters. But by this point, we had in hand all the ingredients needed to prove Gödel's first theorem:

Chapter 12. Take PA as an example. And consider the relation $Prfd(m, n)$, i.e. $Prfd(m, diag(n))$, which holds when m Gödel-numbers a PA proof of the diagonalization of the wff with number n. This relation is primitive recursive, so can be expressed by a formal Σ_1 wff $\mathsf{Prfd}(x, y)$. Now construct the wff $\neg\exists x\,\mathsf{Prfd}(x, y)$.

A moment's reflection shows that a number n satisfies the condition expressed by this open wff if there is no PA proof of the diagonalization of the wff whose Gödel number is n. So diagonalize our wff, and we get a wff G which is true just if there is no PA proof of G (think about it!). It immediately follows that, assuming PA is sound, it can prove neither G nor ¬G.

Inspection shows that G will be a Π_1 wff. And then the argument generalizes to give us our semantic incompleteness theorem:

Theorem 44 *If T is a sound p.r. axiomatized theory whose language contains the language of basic arithmetic, then there will be a true Π_1 sentence G_T such that $T \nvdash \mathsf{G}_T$ and $T \nvdash \neg\mathsf{G}_T$, so T is negation incomplete.*

Chapter 13. Going back to PA, the formal wff $\mathsf{Prfd}(x, y)$ not only expresses but captures $Prfd(m, n)$. Just by relying on the definition of G in terms of $Prfd$ together with the definition of capturing, we can quickly show that the supposition that PA proves G entails that PA is inconsistent.

We can equally quickly show that the supposition that PA proves ¬G implies that PA proves something of the form $\exists x\varphi(x)$ yet also, for each m, proves $\neg\varphi(\overline{m})$, making PA ω-inconsistent.

Then again the argument generalizes to give us our syntactic incompleteness theorem:

Theorem 51 *If T is a consistent p.r. axiomatized theory which contains Q, then there will be a Π_1 sentence G_T such that $T \nvdash \mathsf{G}_T$, and if T is ω-consistent, $T \nvdash \neg\mathsf{G}_T$. Hence, assuming ω-consistency and so consistency, T is negation incomplete.*

Note, the proofs in these last two chapters are not at all difficult. Once you've grasped how to construct a Gödel sentence like G, the rest really is plain sailing. And indeed, it was very important to Gödel himself that the proofs are entirely elementary (and in particular, they call on no infinitary, set-theoretic, resources).

We have scaled our first twin peaks! So where now?

We next excavate the Diagonalization Lemma that can be seen as underlying the Gödelian incompleteness proofs, and then put the Lemma to work to prove two key theorems:

Chapter 14. As in §4.5, use $\mathsf{Prov}_T(x)$ to abbreviate $\exists z\mathsf{Prf}_T(z, x)$. Then a number satisfies the provability predicate $\mathsf{Prov}_T(x)$ iff it Gödel-numbers a T-theorem.

Now, the Gödel sentence G_T is true iff it isn't a T-theorem. Hence $\mathsf{G}_T \leftrightarrow \neg\mathsf{Prov}_T(\ulcorner\mathsf{G}_T\urcorner)$ is true. But there's more. With a bit of manipulation it can be shown that T itself can *prove* $\mathsf{G}_T \leftrightarrow \neg\mathsf{Prov}_T(\ulcorner\mathsf{G}_T\urcorner)$.

And it turns out that this is an instance of something much more general. With the usual sort of assumptions about the theory T, for *any* open wff $\varphi(x)$, there will be a sentence δ such that $\delta \leftrightarrow \varphi(\ulcorner\delta\urcorner)$ will be both true and provable by T. This is the so-called Diagonalization Lemma.

Chapter 15. Tweaking the definition of Prov_T in a way suggested by Rosser, we then get a more complex provability predicate RProv_T. The Diagonalization Lemma will tell us that there is a sentence R_T such that $T \vdash \mathsf{R}_T \leftrightarrow \neg\mathsf{RProv}_T(\ulcorner\mathsf{R}_T\urcorner)$. And (such is Rosser's cunning tweak) we will be able to show that R_T is undecidable by T just by assuming that T is consistent. In this way we can improve on Theorem 51 by dropping the requirement of ω-consistency.

Chapter 16. Suppose for a moment that a theory of the usual kind has an arithmetical predicate T which applies to the code number for a sentence just when that sentence is true – so for any T-sentence φ, $\mathsf{T}(\ulcorner\varphi\urcorner) \leftrightarrow \varphi$.

We could then apply the Diagonalization Lemma again, this time to the negated predicate $\neg\mathsf{T}$, to get a sentence L such that (i) $\mathsf{L} \leftrightarrow \neg\mathsf{T}(\ulcorner\mathsf{L}\urcorner)$. But by assumption, we also have (ii) $\mathsf{T}(\ulcorner\mathsf{L}\urcorner) \leftrightarrow \mathsf{L}$. We immediately get a contradiction: in effect (i) tells us that L is a liar sentence which 'says' it isn't true, and we've hit the Liar Paradox!

This way we prove (one version of) Tarski's Theorem: a theory T can't have a predicate T which fully expresses truth for its own T-sentences (a result which quickly implies incompleteness again).

So the Diagonalization Lemma is important. Our proof is not hard, though it involves an initially non-obvious construction. But is there a simpler and more natural proof? Saul Kripke has claimed to offer one. I present his proof and assess it in a short Appendix at the end of the book.

Continuing, though, with the main thread of chapters, there follow two which further explore the notion of a computable function:

Chapter 17. The primitive recursive functions are, we saw, only a subclass of the effectively computable functions. We now introduce a wider class, the recursive functions. We show that the semantic and syntactic versions of the incompleteness theorem from Chapters 12 and 13 apply not just to p.r. axiomatized theories but also to theories which are, in an obvious sense, recursively axiomatized. But arguably, the recursive functions are *all* the effectively computable total functions, and to be recursively axiomatized is just what it takes to be effectively axiomatized in the informal sense we first met in §2.2. So at this point we can fully align our formally proved incompleteness theorems with the informal versions first introduced in Chapter 3.

Chapter 18. We can't say very much more about computability and decidability in this book. But we can say enough to prove that consistent, recursively axiomatized, theories containing Q aren't decidable, from which it quickly follows that the property of being a theorem of first-order logic is not decidable either. We also touch *very* briefly on the so-called Halting Problem and note its relation to Gödelian incompleteness.

Then, in the remaining group of three chapters, we at long last get to the Second Incompleteness Theorem. In fact, you can if you want jump straight forward from this Interlude to these final chapters (you will only miss the proof of one relevant earlier theorem that we need to use).

Chapter 19. A number satisfies the provability predicate $\mathsf{Prov}_T(\mathsf{x})$ iff it Gödel-numbers a T-theorem. So $\ulcorner\perp\urcorner$, the Gödel number of some suitable absurdity, satisfies the provability predicate $\mathsf{Prov}_T(\mathsf{x})$ iff T is inconsistent. Put $\mathsf{Con}_T =_{\mathrm{def}}$ $\neg\mathsf{Prov}_T(\ulcorner\perp\urcorner)$; then Con_T 'says' that T is consistent.

And now we can state Gödel's Second Theorem: for appropriate formal theories T (consistent, containing enough arithmetic), $T \nvdash \mathsf{Con}_T$ and also $T \nvdash \neg\mathsf{Con}_T$. So we find another undecidable sentence.

Why is this significant? Because it arguably sabotages Hilbert's Programme, as this chapter explains. In headline terms: if a theory can't prove even its own consistency, then we can't hope to use a weaker theory to prove the consistency of a stronger theory.

Chapter 20. The preceding chapter only gives the broadest indication of how the Second Theorem is proved. In this chapter we do better, though we still do not give a full proof (the details are too tedious). We also look at a related result, Löb's Theorem.

Chapter 21. There are some intriguing complications with the Second Theorem. For a start, there are consistent theories that seemingly 'prove their own inconsistency'. And there are, despite the Second Theorem, devious ways in which a theory *can* apparently 'prove its own consistency'. This concluding chapter briefly explores these and related themes.

So onwards, again!

14 The Diagonalization Lemma

As the Interlude announced, we now want to bring to the surface a general result which can be seen as lying behind Gödel's proof of the First Incompleteness Theorem – namely, the Diagonalization Lemma.

In the next chapter, we use the Lemma to prove Rosser's Theorem, a technical improvement on the syntactic First Theorem which allows us to drop the assumption of ω-consistency. Then in Chapter 16 we will put the Lemma to work again, this time to prove Tarski's Theorem, a much deeper result about the 'undefinability of truth'. So the Diagonalization Lemma is important. However, as you will see, its proof is not difficult. In fact, the proofs in this chapter involve little more than choosing the right definitions and then doing some mildly fiddly logical manipulation.

14.1   Two quick reminders

I'll set the scene by restating a couple of definitions.

(a)   First, we pick up an idea we first met in §4.5:

Defn. 16 *Put* $\mathsf{Prov}_T(\mathsf{x}) =_{\mathrm{def}} \exists\mathsf{z}\mathsf{Prf}_T(\mathsf{z},\mathsf{x})$ *(where the quantifier, if necessary, is restricted to run over the natural numbers in the domain).*

Then $\mathsf{Prov}_T(\overline{\mathsf{n}})$, *i.e.* $\exists\mathsf{z}\mathsf{Prf}_T(\mathsf{z},\overline{\mathsf{n}})$, *is* true iff *some number Gödel-numbers a T-proof of the sentence with Gödel-number n, i.e. is true just if the sentence with code number n is a T-theorem. So* Prov_T *is naturally called a* provability predicate.

Therefore $\mathsf{Prov}_T(\overline{\ulcorner\varphi\urcorner})$ is true just when φ is a T-theorem.

(b)   Second, we recall our strengthened version of

Defn. 34 *The theory* T *captures the one-place function* f *by the open wff* $\psi(\mathsf{x},\mathsf{y})$ *iff, for any* m, n,

     i. *if* $f(m) = n$, *then* $T \vdash \psi(\overline{\mathsf{m}},\overline{\mathsf{n}})$,
    ii′. $T \vdash \exists!\mathsf{y}\psi(\overline{\mathsf{m}},\mathsf{y})$.

And this time let's note that (i) and (ii′) together obviously imply

    (*) *if* $f(m) = n$, *then* $T \vdash \forall\mathsf{x}(\psi(\overline{\mathsf{m}},\mathsf{x}) \leftrightarrow \mathsf{x} = \overline{\mathsf{n}})$,

a fact we'll need in a moment.

14.2 Proving $G_T \leftrightarrow \neg Prov_T(\ulcorner G_T \urcorner)$

Recall: our Gödel sentence G_T is the diagonalization of $U(y) =_{def} \neg \exists x \, Prfd_T(x, y)$. And we have shown that G_T is true if and only if unprovable-in-T.

Now, given what we've just said about $Prov_T$, that key fact about G_T can be *expressed* inside T itself, by the wff $G_T \leftrightarrow \neg Prov_T(\ulcorner G_T \urcorner)$. But T doesn't just express this fact – the theory can *prove* it too:

Theorem 53 *If T is p.r. axiomatized and contains* Q, $T \vdash G_T \leftrightarrow \neg Prov_T(\ulcorner G_T \urcorner)$.

The proof is elementary, just appealing to our definitions and to simple logical manipulations:

Proof Dropping subscripts, $Prfd(m, n)$ holds iff $Prf(m, diag(n))$. As noted in Theorem 40, we can therefore fix on the following canonical definition:

$$Prfd(x, y) =_{def} \exists z(Prf(x, z) \wedge Diag(y, z)).$$

Next, since diagonalizing U yields G, we have $diag(\ulcorner U \urcorner) = \ulcorner G \urcorner$ (just by the definition of *diag*). Since Diag captures *diag*, it follows – by the remark (*) of the previous section – that

(**) $T \vdash \forall z(Diag(\ulcorner U \urcorner, z) \leftrightarrow z = \ulcorner G \urcorner)$.

Then, arguing inside theory T, we have:

$G \leftrightarrow$	$\neg \exists x \, Prfd(x, \ulcorner U \urcorner)$	(by defn of G)
$\leftrightarrow$	$\neg \exists x \exists z(Prf(x, z) \wedge Diag(\ulcorner U \urcorner, z))$	(by defn of Prfd)
$\leftrightarrow$	$\forall z(Diag(\ulcorner U \urcorner, z) \to \neg \exists x \, Prf(x, z))$	(by logical manipulation)
$\leftrightarrow$	$\forall z(Diag(\ulcorner U \urcorner, z) \to \neg Prov(z))$	(by defn of Prov)
$\leftrightarrow$	$\forall z(z = \ulcorner G \urcorner \to \neg Prov(z))$	(using (**))
$\leftrightarrow$	$\neg Prov(\ulcorner G \urcorner)$	(by logical manipulation) ☒

14.3 The syntactic First Theorem again

(a)   The *truth* of $G_T \leftrightarrow \neg Prov_T(\ulcorner G_T \urcorner)$ easily gives us the semantic version of the First Theorem. The *provability-in-T* of that sentence gives us the syntactic version.

And in fact, have a more general result:

Theorem 54 *Suppose T is p.r. axiomatized, contains* Q, *and suppose some sentence γ is such that $T \vdash \gamma \leftrightarrow \neg Prov_T(\ulcorner \gamma \urcorner)$. Then (i) if T is consistent, $T \nvdash \gamma$. And (ii) if T is ω-consistent, $T \nvdash \neg \gamma$.*

So not just G_T but any sentence γ which is provably equivalent to a statement of its own unprovability will be undecidable in T.

Again dropping subscript 'T's for the moment, we can argue like this, *very* closely following our proofs of Theorems 46 and 47:

Proof (for i) Suppose $T \vdash \gamma$. Then, since $T \vdash \gamma \leftrightarrow \neg Prov_T(\ulcorner \gamma \urcorner)$, we have $T \vdash \neg Prov(\ulcorner \gamma \urcorner)$. But if there *is* a proof of γ, then for some m, $Prf(m, \ulcorner \gamma \urcorner)$, so

$T \vdash \mathsf{Prf}(\overline{m}, \ulcorner\gamma\urcorner)$, since T captures Prf by Prf. Hence $T \vdash \exists x\, \mathsf{Prf}(x, \ulcorner\gamma\urcorner)$, i.e. we also have $T \vdash \mathsf{Prov}(\ulcorner\gamma\urcorner)$, making T inconsistent. So if T is consistent, $T \nvdash \gamma$. ⊠

Proof (for ii) Suppose T is consistent and $T \vdash \neg\gamma$. Then, since $T \vdash \gamma \leftrightarrow \neg\mathsf{Prov}_T(\ulcorner\gamma\urcorner)$, we have $T \vdash \mathsf{Prov}(\ulcorner\gamma\urcorner)$, i.e. $T \vdash \exists x\, \mathsf{Prf}(x, \ulcorner\gamma\urcorner)$. Also, given T is consistent and proves $\neg\gamma$, there is no proof of γ, i.e. for every m, not-$Prf(m, \ulcorner\gamma\urcorner)$, whence for every m, $T \vdash \neg\mathsf{Prf}(\overline{m}, \ulcorner\gamma\urcorner)$. So we have a $\psi(x)$ such that T proves $\exists x\psi(x)$ while it refutes each instance $\psi(\overline{m})$.

Hence, if T is consistent and $T \vdash \neg\gamma$, T is ω-inconsistent. So if T is ω-consistent (and hence consistent), $T \nvdash \neg\gamma$. ⊠

(b)   A bit of standard jargon. By a mild abuse of mathematical terminology, we say:

Defn. 48 *Suppose that φ is a T-wff with one free variable. Then δ is said to be a* fixed point *for φ if and only if $T \vdash \delta \leftrightarrow \varphi(\ulcorner\delta\urcorner)$.*

So we can prove the syntactic First Theorem in two steps. First prove that there is fixed point for $\neg\mathsf{Prov}_T(x)$ (showing that G_T fits the bill). Then prove that any fixed point for $\neg\mathsf{Prov}_T(x)$ is formally undecidable in T.[1]

14.4   The Diagonalization Lemma

We've seen that the wff $\neg\mathsf{Prov}(x)$ has a fixed point in the sense just explained. But there is nothing special about this case. In fact we have the following very general *Diagonalization Lemma*. Assume that T is a p.r. axiomatized theory, that a normal Gödel-numbering scheme is in play, and that φ is a one-place open sentence of T's language. Then:

Theorem 55 *(i) If T contains the language L_A (the language of Q) there is a sentence δ such that $\delta \leftrightarrow \varphi(\ulcorner\delta\urcorner)$ is true, and moreover*

(ii) if T contains Q, then $T \vdash \delta \leftrightarrow \varphi(\ulcorner\delta\urcorner)$.

Think what was involved in constructing the fixed point G_T. There were two key steps. We first constructed the wff Prfd by combining Prf with Diag using an existential quantifier. Then we diagonalized. We'll dance the same two-step again:

(1)   At our first step we put $\alpha =_{\mathrm{def}} \exists z(\mathsf{Diag}(x, z) \wedge \varphi(z))$.

(2)   Then, second step, we let δ be the diagonalization of *that*. So,

$$\delta =_{\mathrm{def}} \exists z(\mathsf{Diag}(\ulcorner\alpha\urcorner, z) \wedge \varphi(z)).$$

---

[1]Warning. Some authors call *any* fixed point for *any* predicate $\neg\mathsf{Prov}_T$ built from some wff which captures Prf_T a Gödel sentence for T. That's reasonable enough, since a Gödel sentence in this sense is formally undecidable. However, this wide usage is still best avoided. Because not everything that is true of canonical Gödel sentences built up in the way we introduced in the preceding chapters will be true of Gödel sentences in this new wider sense. For example, Gödel sentences in the wide sense needn't be 'about' their own unprovability (see Theorem 73). And they can be even false, when we are dealing with an unsound theory.

We now show that (i) and (ii) in our Theorem are true for this δ (for note that Diag is available, given our assumptions). We just need to invoke definitions and infer easy logical consequences.

Proof (for i) Because diagonalizing α yields δ, by definition $diag(\ulcorner\alpha\urcorner) = \ulcorner\delta\urcorner$. Since Diag expresses $diag$, $\mathsf{Diag}(\ulcorner\alpha\urcorner, \ulcorner\delta\urcorner)$ will therefore be true; and in fact, $\mathsf{Diag}(\ulcorner\alpha\urcorner, \mathsf{z})$ is *only* satisfied by $\ulcorner\delta\urcorner$.

This implies that $\exists\mathsf{z}(\mathsf{Diag}(\ulcorner\alpha\urcorner, \mathsf{z}) \wedge \varphi(\mathsf{z}))$ is true if and only if $\ulcorner\delta\urcorner$ satisfies $\varphi(\mathsf{z})$. In other words, δ is true if and only if $\varphi(\ulcorner\delta\urcorner)$ is true. $\boxtimes$

Proof (for ii) Since Diag captures $diag$ in T, (*) from the beginning of the chapter, §14.1, tells us that if $diag(\ulcorner\alpha\urcorner) = \ulcorner\delta\urcorner$, then $T \vdash \forall\mathsf{z}(\mathsf{Diag}(\ulcorner\alpha\urcorner, \mathsf{z}) \leftrightarrow \mathsf{z} = \ulcorner\delta\urcorner)$.

But, as we just noted, $diag(\ulcorner\alpha\urcorner) = \ulcorner\delta\urcorner$. Hence we can conclude that, indeed, $T \vdash \forall\mathsf{z}(\mathsf{Diag}(\ulcorner\alpha\urcorner, \mathsf{z}) \leftrightarrow \mathsf{z} = \ulcorner\delta\urcorner)$.

So T proves the equivalence of $\mathsf{Diag}(\ulcorner\alpha\urcorner, \mathsf{z})$ and $\mathsf{z} = \ulcorner\delta\urcorner$. Therefore T can also prove the equivalence of δ, i.e. $\exists\mathsf{z}(\mathsf{Diag}(\ulcorner\alpha\urcorner, \mathsf{z}) \wedge \varphi(\mathsf{z}))$, with $\exists\mathsf{z}(\mathsf{z} = \ulcorner\delta\urcorner \wedge \varphi(\mathsf{z}))$. But the latter is in turn trivially equivalent to $\varphi(\ulcorner\delta\urcorner)$.

Hence $T \vdash \delta \leftrightarrow \varphi(\ulcorner\delta\urcorner)$. $\boxtimes$

We will put this result to work in the next two chapters.[2]

---

[2] You could at this point look at the Appendix on a variant Diagonalization Lemma due to Kripke, which is advertised as involving a more natural construction.

15 Rosser's Theorem

One half of our syntactic First Theorem requires the assumption that we are dealing with a theory T which is not only consistent but also ω-consistent. We can now improve on this. Following J. Barkley Rosser (in 1936), we can construct a sentence R_T such that we only need to assume T is plain consistent in showing that R_T is formally undecidable. This is a technical result you really ought to know about and we can use the Diagonalization Lemma to prove it. But the proof-details are a bit messy; you really are allowed to skim or skip them!

15.1   Tweaking the provability predicate, Rosser-style

Assume that a Gödel-numbering scheme is in place, and define $\overline{Prf}_T(m, n)$ as holding when m is the number of a T-proof of the *negation* of the wff with number n. This relation is obviously p.r. given that Prf_T is. Hence, assuming T has the usual properties, $\overline{Prf}$ can be expressed and captured by a wff $\overline{\mathsf{Prf}}_T(\mathsf{x}, \mathsf{y})$. Now consider:

Defn. 49 $\mathsf{RPrf}_T(\mathsf{x}, \mathsf{y}) =_{\text{def}} \mathsf{Prf}_T(\mathsf{x}, \mathsf{y}) \wedge (\forall \mathsf{w} \leq \mathsf{x})\neg\overline{\mathsf{Prf}}_T(\mathsf{w}, \mathsf{y})$.

A pair of numbers m, n will satisfy this complex predicate if m numbers the proof of a wff with number n, and no smaller number numbers the proof of that wff's negation. Assuming T is consistent, then, a pair of numbers m, n will satisfy this complex predicate just if m numbers the proof of a wff with number n. So RPrf_T still expresses the relation Prf_T.

But there's more. This tweaked predicate RPrf_T in fact also captures the relation Prf_T, again assuming T is consistent. Why?

(1) Suppose, supressing subscripts, $Prf(m, n)$. Then (i) $T \vdash \mathsf{Prf}(\overline{m}, \overline{n})$.

   Moreover, given that T is consistent, since the wff with number n has a proof, no number numbers a proof of the negation of that wff. Which means that $\overline{Prf}(0, n)$, $\overline{Prf}(1, n)$, $\overline{Prf}(3, n)$, ... $\overline{Prf}(m, n)$ are all *false*.

   Hence, because $\overline{\mathsf{Prf}}$ captures $\overline{Prf}$, we have each of $T \vdash \neg\overline{\mathsf{Prf}}(\overline{0}, \overline{n})$, $T \vdash \neg\overline{\mathsf{Prf}}(\overline{1}, \overline{n})$, $T \vdash \neg\overline{\mathsf{Prf}}(\overline{2}, \overline{n})$, ..., $T \vdash \neg\overline{\mathsf{Prf}}(\overline{m}, \overline{n})$.

   Therefore, we will also have (ii) $T \vdash (\forall \mathsf{w} \leq \overline{m})\neg\overline{\mathsf{Prf}}(\mathsf{w}, \overline{n})$ – since Q and hence T 'knows' about bounded quantifiers (see §8.1).

   So from (i) and (ii) we have $T \vdash \mathsf{RPrf}(\overline{m}, \overline{n})$.

(2) Suppose not-$Prf(m, n)$. Then $T \vdash \neg\mathsf{Prf}(\overline{m}, \overline{n})$, so $T \vdash \neg\mathsf{RPrf}(\overline{m}, \overline{n})$.

And now we can define *the Rosser provability predicate* as follows:

Defn. 50 $\mathsf{RProv}_T(x) =_{\text{def}} \exists v \mathsf{RPrf}(v, x)$.

Then a sentence is Rosser-provable in T and its g.n. satisfies the Rosser provability predicate if and only if it has a proof (in the ordinary sense) and there's no smaller-numbered proof of its negation.

Note for use in a moment that it follows immediately from the definition that

$$\neg\mathsf{RProv}_T(x) \leftrightarrow \forall x(\mathsf{Prf}_T(x, y) \to (\exists w \leq x)\overline{\mathsf{Prf}}_T(w, y)).$$

15.2 Rosser's Theorem

Now we apply the Diagonalization Lemma, not to the negation of a regular provability predicate, but to the negation of the Rosser provability predicate. The Lemma then tells us,

Theorem 56 *Given that T is p.r. axiomatized and contains Q, then there is a sentence R_T such that $T \vdash \mathsf{R}_T \leftrightarrow \neg\mathsf{RProv}_T(\ulcorner\mathsf{R}_T\urcorner)$.*

We call such a sentence R_T a Rosser sentence for T. Then we have this key result:

Theorem 57 *Suppose T is a consistent p.r. axiomatized theory which contains Q and let ρ be any fixed point for $\neg\mathsf{RProv}_T(x)$. Then (i) $T \nvdash \rho$ and (ii) $T \nvdash \neg\rho$.*

Proof (of i) Just follow the proof of (i) for Theorem 54 – simply change $\mathsf{Prf}/\mathsf{Prov}$ to $\mathsf{RPrf}/\mathsf{RProv}$, and use the fact that RPrf also captures *Prf*. ⊠

Proof (of ii, messy!) Suppose, for reductio, that $T \vdash \neg\rho$. We argue in two stages.

(1) Since by assumption $T \vdash \rho \leftrightarrow \neg\mathsf{RProv}_T(\ulcorner\rho\urcorner)$, our supposition implies $T \vdash \mathsf{RProv}(\ulcorner\rho\urcorner)$.

(2) Because we are supposing $\neg\rho$ is provable, for some m, $\overline{Prf}(m, \ulcorner\rho\urcorner)$, so we have (a) $T \vdash \overline{\mathsf{Prf}}(\overline{m}, \ulcorner\rho\urcorner)$.

Also, since T is consistent, ρ is unprovable, so for all n, not-$Prf(n, \ulcorner\rho\urcorner)$. Hence, in particular, for each n up to m, $T \vdash \neg\mathsf{Prf}(\overline{n}, \ulcorner\rho\urcorner)$. From which we get $T \vdash (\forall v \leq \overline{m})\neg\mathsf{Prf}(v, \ulcorner\rho\urcorner)$ since T knows about bounded quantifiers. Equivalently, (b) $T \vdash \forall v(\mathsf{Prf}(v, \ulcorner\rho\urcorner) \to \overline{m} \leq v)$.

Combining (a) and (b), it immediately follows that $T \vdash \forall v(\mathsf{Prf}(v, \ulcorner\rho\urcorner) \to (\overline{m} \leq v \wedge \overline{\mathsf{Prf}}(\overline{m}, \ulcorner\rho\urcorner)))$. Hence $T \vdash \forall v(\mathsf{Prf}(v, \ulcorner\rho\urcorner) \to (\exists x \leq v)\overline{\mathsf{Prf}}(x, \ulcorner\rho\urcorner)))$. Hence $T \vdash \neg\mathsf{RProv}(\ulcorner\rho\urcorner)$.

But now we have shown both (1) $T \vdash \mathsf{RProv}(\ulcorner\rho\urcorner)$ and (2) $T \vdash \neg\mathsf{RProv}(\ulcorner\rho\urcorner)$, contradicting T's consistency. Which refutes our supposition that $T \vdash \neg\rho$. ⊠

Putting our last two theorems together, we get the Rosser's Theorem:

Theorem 58 *If T is a consistent, p.r. axiomatized theory which contains Q, then there is sentence R_T such that, if T is consistent then $T \nvdash \mathsf{R}_T$ and $T \nvdash \neg\mathsf{R}_T$.*

With yet more work, we can beef up the proof idea to show that in fact (as with Gödel's original proof) we can find a Π_1 sentence which is undecidable so long as T is consistent. But we won't do that here.

16 Tarski's Theorem

The last chapter involved a bit of trickery in defining the Rosser provability predicate, and then we did some unexciting manipulations to prove Rosser's Theorem. You may well have skimmed over the details. In this chapter, however, the basic proofs are as simple as can be, given the Diagonalization Lemma.

The Lemma, recall, has two parts. Take a suitable theory T. Then if φ is a one-place open sentence of T's language, then there is a sentence δ such that (i) $\delta \leftrightarrow \varphi(\ulcorner \delta \urcorner)$ is true and (ii) $T \vdash \delta \leftrightarrow \varphi(\ulcorner \delta \urcorner)$.

This chapter applies each part of the Lemma, to give a pair of results about truth that are usually packaged together as *Tarski's Theorem*. We arrive at the deep contrast between the notion of truth and the notion of provability which Gödel saw as underlying the incompleteness phenomenon.

16.1   Truth predicates and truth definitions

Start from a familiar thought. 'Snow is white' is true iff snow *is* white. Likewise for all other sensible replacements for 'snow is white'. In sum, we can endorse every sensible instance of *'φ' is true iff φ*.[1] And that's because of the meaning of 'is true'. Generalizing, let's say that an informal predicate *Tr* serves as a *truth predicate* iff every instance of *Tr('φ') iff φ* is likewise true.

How can we add an analogous truth predicate to an interpreted formal language L which contains the language of basic arithmetic (as in Defn. 11)? Such a language will in general not have quotation marks or the like available; however, we can arithmetize syntax and use code numbers to refer to wffs. Assume that we have fixed on some normal scheme for Gödel numbering L-wffs. Then we can define a corresponding numerical property *True* as follows:

> *True(n)* is true iff n is the g.n. of a true sentence of L.

Now suppose, just suppose, we introduce some expression $\mathsf{T}(\mathsf{x})$ with one free variable which is so defined as to *express* this numerical property *True*. And – allowing for the possibility that we've had to extend L in introducing such an expression – let L^* be the result of adding a new wff $\mathsf{T}(\mathsf{x})$ to our initial language L if necessary. (So for the moment, we leave it open whether L^* is just L, which it would be if a suitable $\mathsf{T}(\mathsf{x})$ is already definable from L's resources.) Then, just

---

[1] Forgive the casualness with the quotation marks. Nothing will depend on this!

by the stipulated definitions of *True* and of T, we have the following for any sentence φ of the original language L:

φ is true iff $True(\ulcorner\varphi\urcorner)$ iff $T(\ulcorner\varphi\urcorner)$ is true.

Hence, for any L-sentence φ, every corresponding 'T-biconditional'

$$T(\ulcorner\varphi\urcorner) \leftrightarrow \varphi$$

is true. Which motivates our first main definition:

Defn. 51 *An open L^*-wff* T(x) *(where L^* includes L) is a* truth predicate *for L iff for every L-sentence φ, $T(\ulcorner\varphi\urcorner) \leftrightarrow \varphi$ is true.*

And here's a companion definition:

Defn. 52 *A theory Θ (with language L^* which includes L) is a* formal truth theory *for L iff it provides an L^*-wff* T(x) *such that $\Theta \vdash T(\ulcorner\varphi\urcorner) \leftrightarrow \varphi$ for every L-sentence φ.*

Conventionally, a truth theory for L is also called a *definition of truth* for L.

In sum, a truth predicate T for L is a predicate of a perhaps extended language that applies to (the Gödel numbers for) exactly the true L-sentences, and so *expresses* truth. And a truth theory for L is a theory built in a perhaps extended language that *proves* all the T-biconditionals for L sentences.

So far, however, that's just a sequence of (natural enough) definitions. Now for our first big result.

16.2 The undefinability of truth

Suppose T is an arithmetical theory with language L. The question arises: could T be competent to define truth *for its own language* (in other words, can T already encompass a truth theory for L)? And an answer is immediate – call this Tarski's Theorem on the undefinability of truth:

Theorem 59 *No consistent p.r. axiomatized theory T which contains Q can define truth for its own language.*

Proof Assume T defines truth for L using an open L-sentence T(x). Then

(1) For all L-sentences φ, $T \vdash T(\ulcorner\varphi\urcorner) \leftrightarrow \varphi$.

Since T has the right properties, part (ii) of the Diagonalization Lemma applies. Therefore we can apply the Lemma in particular to $\neg T(x)$, so there must be some sentence L such that

(2) $T \vdash L \leftrightarrow \neg T(\ulcorner L\urcorner)$.

According to T then, L is a Liar sentence, which (as it were) says that it is false! But, by our initial assumption that (1) holds, we also have

(3) $T \vdash T(\ulcorner L\urcorner) \leftrightarrow L$.

But (2) and (3) together entail that T is inconsistent, contrary to hypothesis. So our assumption must be wrong: T can't define truth for its own language. ⊠

16.3   Curry's Paradox and Tarski's Theorem

(a)   Our proof that a theory T (satisfying the usual conditions) can't define truth for its own language was essentially this: if T could define truth, it would get entangled with the Liar Paradox. It is fun and instructive to give another proof of the same theorem: this time, we show that, if T could define truth for its own language, it would get entangled with Curry's Paradox. But what's that?

(b)   Start from this thought: we seemingly can construct a sentence that says e.g. 'if this sentence is true, then the moon is made of green cheese' (after all, haven't I just constructed one?). Let's symbolize the whole sentence by δ, and let's put ψ for 'the moon is made of green cheese'.

So by construction, δ supposedly holds exactly when, if δ, then ψ. And now we can argue as follows – just using the familiar inference rules for the conditional and biconditional:

(1)	$\delta \leftrightarrow (\delta \to \psi)$	By definition of δ (?)
(2)	δ	Supposition
(3)	$\delta \to \psi$	From 1, 2
(4)	ψ	From 2, 3
(5)	$\delta \to \psi$	Conditional proof from 2 to 4
(6)	δ	From 1, 5
(7)	ψ	From 5, 6

So we have proved that the moon is made of green cheese!

This sort of argument was already known to medieval logicians. But it was rediscovered by Haskell Curry in 1942, and these days – in one form or another – is known as *Curry's Paradox*. Why 'paradox'? Because, at first sight, truths of the kind (1) seem to be available to us, assuming we can construct self-referential sentences at all (and these are surely often harmless, as in 'This sentence contains five words'). Yet evidently something has gone wrong.

Is there anything problematic *after* the starting assumption at line (1)? Seemingly not: for we have only used intuitively secure inference moves. We have simply derived one-way conditionals from a biconditional, and then reasoned about conditionals using modus ponens and conditional proof. So, the blame surely has to fall on the original assumption that there is such a δ as makes (1) hold.

But what is wrong with this assumption? Let's dig deeper. We took a sentence 'if this sentence is true, then the moon is made of green cheese', symbolized the whole conditional as δ, symbolized the consequent of the conditional as ψ, and then asserted that we have

(1) $\delta \leftrightarrow (\delta \to \psi)$.

However, that elided two steps. Strictly speaking, if δ symbolises the whole conditional, the antecedent of the conditional should be $True(\delta)$, and our initial equivalence will then be

(E) $\delta \leftrightarrow (True(\delta) \to \psi)$

And then we need to appeal to the equivalence

(T) $True(\delta) \leftrightarrow \delta$

to arrive at the troublesome (1). So which of (E) and (T) should we give up?

(c)   We are not going to tackle this question in general. But let's consider the argument in a formal context. So now assume that both

(I) T is a p.r. axiomatized theory which contains Q;
(II) T can define truth for its own language using the truth predicate $\mathsf{T}(x)$.

Then we can argue as follows.

Take a sentence ψ of T, and consider the open sentence $\mathsf{T}(x) \to \psi$. Given (I), we have the Diagonalization Lemma, so corresponding to (E) there will indeed be some δ such that

(1) $T \vdash \delta \leftrightarrow (\mathsf{T}(\overline{\ulcorner \delta \urcorner}) \to \psi)$

And then since, by (II), T can define truth, we have

(2) $T \vdash \mathsf{T}(\overline{\ulcorner \delta \urcorner}) \leftrightarrow \delta$

corresponding to the informal (T) above. And now we can proceed as follows

(3) $T \vdash \delta \leftrightarrow (\delta \to \psi)$ From 1, 2, using propositional logic in T
(4) $T \vdash \psi$ Arguing in T from 3 as for Curry's Paradox

But ψ was arbitrarily chosen. So T can prove any sentence and hence is inconsistent. So if T is consistent and our assumption (I) about T is true, then (II) is false. Which is just Tarski's Theorem 59 again.

16.4   The inexpressibility of truth

Our first Tarskian theorem puts limits on what a nice theory can *prove* about truth. But we can go further: there are limits on what a theory's language can even *express* about truth.

Consider our old friend L_A for the moment, and suppose that there is an L_A truth predicate T_A that expresses the corresponding truth property $True_A$. Then part (i) of the Diagonalization Lemma, applies to $\neg\mathsf{T}_A(x)$ (for the proof of this semantic part depended only on the fact that we were dealing with a language which contains the language of basic arithmetic). So in particular, there will be some L_A sentence L such that

(1) $\mathsf{L} \leftrightarrow \neg\mathsf{T}_A(\overline{\ulcorner \mathsf{L} \urcorner})$.

is true. But, by the assumption that T_A is a truth predicate for L_A,

(2) $\mathsf{T}_A(\overline{\ulcorner \mathsf{L} \urcorner}) \leftrightarrow \mathsf{L}$

must be true too. (1) and (2) immediately lead to contradiction again. Therefore our supposition that T_A expresses the property of being a true L_A sentence has to be rejected.

The argument generalizes. Take any language L which includes the language of basic arithmetic, so that the first part of the Diagonalization Lemma is provable.

Then by the same argument we get Tarski's Theorem limiting the expressibility of truth:

Theorem 60 *No predicate of a language L which includes the language of basic arithmetic can express the numerical property* $True_L$ *(i.e. the property of numbering a truth of L).*

16.5   The Master Argument for incompleteness?

Our second Tarskian result tells us that while you can express *syntactic* properties of a sufficiently rich formal theory of arithmetic (like provability) inside that theory via Gödel numbering, you can't express some key *semantic* properties (like truth) inside the same theory. And this points to a particularly illuminating take on the argument for incompleteness.

For example: PA-provability *is* expressible in L_A (by a provability predicate Prov), but truth-in-L_A *isn't* expressible in L_A. Hence truth in L_A isn't provability in PA. So assuming that PA is sound so that everything provable in it is true, this means that there must be truths of L_A which PA can't prove. Similarly, of course, for other nice theories.

And in a way, we might well take this to be *the* Master Argument for incompleteness, revealing the true roots of the phenomenon. Gödel himself wrote (in response to a query)

> I think the theorem of mine that von Neumann refers to is ... that a complete epistemological description of a language A cannot be given in the same language A, because the concept of truth of sentences in A cannot be defined in A. *It is this theorem which is the true reason for the existence of undecidable propositions in the formal systems containing arithmetic.* I did not, however, formulate it explicitly in my paper of 1931 but only in my Princeton lectures of 1934. The same theorem was proved by Tarski in his paper on the concept of truth.[2]

In sum, as we emphasized before, arithmetical truth and provability in this or that formal system must peel apart.

"How does that statement of Gödel's square with the importance that he placed on the syntactic version of the First Theorem?" Well, Gödel himself was a realist about mathematics (certainly he was in his later philosophical writings; and he claimed that this had been his view all along). So, the story goes, *he* believed in the real existence of mathematical entities, and believed that our theories (at least aim to) deliver truths about them. But that wasn't the dominant belief among those around him concerned with foundational matters. As I noted in §3.2, for various reasons the very idea of truth-in-mathematics was under some suspicion at the time. So even if semantic notions were at the

---

[2]Gödel's letter is quoted in Solomon Feferman's very interesting 'Kurt Gödel: conviction and caution', reprinted in Feferman's *In the Light of Logic* (OUP 1998).

root of Gödel's insight, it was extremely important for him – given the intended audience – to show that you don't need to deploy those semantic notions to prove incompleteness.

"But does this mean that, if *we* are happy enough with the use of semantic premisses, we needn't worry about tackling the complications of proving the syntactic version of the First Theorem?" No. For a start, as we will see, it is reflection on the syntactic First Theorem which will take us to the Second Theorem.

17 Recursive functions

We saw in Chapter 9 that the primitive recursive functions are a centrally important subclass, but only a subclass, of the effectively computable functions. In this chapter we'll say something about the more inclusive class of recursive functions (but only just enough for our present purposes).

Now, we remarked in §12.6 on the gap between the semantic incompleteness theorem as originally announced and the result we have actually proved. The first, Theorem 1, applies to effectively axiomatized theories; the second, Theorem 44, applies only to p.r. axiomatized theories. One payoff from the discussions in this chapter is that – with a bit of help from what's called Church's Thesis – we will be able to bridge the gap between those theorems.

Similarly, we can bridge the gap noted in §13.5 between the originally announced syntactic incompleteness theorem, Theorem 2, and the version we have proved, Theorem 51.

17.1 Recursive functions

(a) The primitive recursive functions are the numerical functions that can be computed without open-ended searches. So in order to get a more inclusive class of computable functions, let's make the obvious next move and now allow such searches.

The standard way of implementing this move is by introducing a *least number* or *minimization* operator, to be explained very shortly. But note: the search for the least number satisfying a certain condition may never terminate. Therefore specifying a computable function using a least number search may in general result in a *partial* function, i.e. one which isn't defined for some inputs. We still want, however, to characterize a class of effectively computable *total* functions. There are slightly different ways of getting there, which in the end come to the same. I choose the simplest and most direct route.

(b) Let's start with a motivating base case. Suppose that (i) $g(x, y)$ is a total computable function. Suppose also that (ii) for all x there is one or more value of y such that $g(x, y) = 0$. Then we can define a function f as follows: $f(x)$ *is the least value of y such that $g(x, y) = 0$.* And so defined, f is again a total computable function.

Why so? Because you can set out to determine the value of f for a particular input x by computing $g(x, 0)$, $g(x, 1)$, $g(x, 2)$, $\ldots$, in turn. You can do this since by assumption (i) $g(x, y)$ can always be effectively computed. Carry on until you find the first value of y such that $g(x, y) = 0$, and output that value of y. Our assumption (ii) ensures that this open-ended search routine does always terminate, so giving an output for any given input x.

Hence in this case f is indeed a total computable function, one which can be evaluated using a certain kind of looping computation (though note that, unlike the case of a basic 'for' loop, the actual number of iterations round the loop which are needed will this time not be explicitly fixed in advance).

(c)   Now we generalize.

Defn. 53 *A $(k + 1)$-place function g is* regular *iff for all k-tuples $\vec{x}$ there is a value of y such that $g(\vec{x}, y) = 0$.*

Defn. 54 *The k-place function f is defined by* minimization *from the $(k + 1)$-place function g iff $f(\vec{x})$ is the least value of y such that $g(\vec{x}, y) = 0$.*[1]

We can immediately see that, as before, if g is an effectively computable total function *and* is regular, and f is defined from it by minimization, then f too is an effectively computable total function.

(d)   We now we have what we need to define the recursive functions.

The primitive recursive functions, recall, are those that can be defined from our stock of initial functions by repeated applications of composition and primitive recursion. Similarly, the recursive functions are those that can be defined from the same initial functions by repeated applications of composition, primitive recursion, and minimization of regular functions.

In other words,

Defn. 55 *The recursive functions are as follows:*

(1) The successor function S, zero function Z, and all the identity functions I_i^k are recursive;

(2) if f can be defined from the recursive functions g and h by composition, substituting g into h, then f is recursive;

(3) if f can be defined from the recursive functions g and h by primitive recursion, then f is recursive;

(4) if g is recursive and regular, and f is defined by minimization from g, then f is recursive;

(5) nothing else is a recursive function.

So the recursive functions include the primitive recursive ones and, given what we said about minimization of regular functions, they are again always effectively computable total functions.

---

[1] Fine print: this is a cruder definition than you will usually find elsewhere, which is given in the rather different context of defining *partial* recursive functions. But our simple version works for us because we are only going to care about minimizing regular total functions.

113

As you would expect, we will also say a property or relation is recursive if and only if its characteristic function is recursive; so recursive properties and relations are effectively decidable.

17.2 Expressing and capturing recursive functions

If the one-place function f is defined from g by minimization, then $f(m) = n$ iff both $g(m, n) = 0$ and for all $k < n$, $g(m, k) = z$ where $z \neq 0$. So suppose that the L_A-wff $\mathsf{G}(\mathsf{x}, \mathsf{y}, \mathsf{z})$ expresses g; if we put

$$\mathsf{F}(\mathsf{x}, \mathsf{y}) =_{\mathrm{def}} \mathsf{G}(\mathsf{x}, \mathsf{y}, 0) \wedge (\forall \mathsf{k} \leq \mathsf{y})(\mathsf{k} \neq \mathsf{y} \rightarrow \exists \mathsf{z}(\mathsf{G}(\mathsf{x}, \mathsf{k}, \mathsf{z}) \wedge \mathsf{z} \neq 0))$$

then evidently F expresses f in L_A. And by inspection, if G is Σ_1, so is F.

The point generalizes to a many-place function f defined by regular minimization from a many-place g. The same simple construction turns a Σ_1 wff expressing g into one expressing f.

Let's now put this simple observation to work. Recall, every primitive recursive function f has at least one definition chain: in other words, there will be a sequence of functions $f_0, f_1, f_2, \ldots, f_k$ where each f_j is either an initial function or is defined from previous functions in the sequence by composition or recursion, and $f_k = f$. We then showed in §10.3 how we can build up a canonical Σ_1 wff expressing f by tracking that definition chain step by step. Exactly similarly, every recursive function f also has at least one definition chain where each f_j in the chain is either an initial function or is defined from previous functions in the sequence by composition, recursion, or minimization. And, just as before, we can build up a canonical Σ_1 wff expressing the recursive function f by tracking its definition chain, because we now know how to handle the definitions by minimization.

That's easy. And with rather more work we can also show that if G captures g, then F (as above) captures the function f defined from g by minimization. Hence we can similarly strengthen our proof that (a theory including) Q can capture every primitive recursive function to show that it can capture every recursive function. We needn't go into the distracting details of the proof, but will just appeal to the result when needed.

17.3 Fast growing functions

(a) Theorem 29 tells us that there are computable functions which aren't primitive recursive; we proved that by constructing a rather artificial example, using a diagonal argument. Can we give a more natural example of a computable, recursive-but-not-primitive-recursive, function?

Consider the following two-place p.r. functions: f_1 is *sum* (repeated applications of the successor function); f_2 is *product* (repeated sums); f_3 is *exponentiation* (repeated products). As their respective arguments grow, the value of f_1 of course grows comparatively slowly, f_2 grows faster, f_3 faster still.

This sequence of p.r. functions can obviously be continued. Next comes f_4, the *super-exponential*, defined by repeated exponentiation:

$$x \Uparrow 0 = x$$
$$x \Uparrow Sy = x^{x \Uparrow y}.$$

Thus, for example, $3 \Uparrow 4$ is $3^{3^{3^{3^3}}}$ with a 'tower' of four exponents. Similarly, we can define f_5 (super-duper-exponentiation, i.e. repeated super-exponentiation), f_6 (repeated super-duper-exponentiation), and so on. The values of these p.r. functions grow faster and faster as their arguments are increased.

So now let's define $A(x) = f_x(x, x)$. The value of $A(x)$ evidently grows quite *explosively*, running away ever faster as x increases.

Now, we can show that for any given p.r. function $g(x)$ there is an n such that its rate of growth as x increases is no faster than that of $f_n(x, x)$.[2] But obviously, $f_x(x, x)$ eventually grows faster than any particular $f_n(x, x)$, when $x > n$. Hence the function A can't itself be primitive recursive. Yet $A(x)$ is effectively computable – proceed to the definition of f_x, and compute its value for the inputs x, x. Moreover, with just a bit of work, A can be directly shown to be recursive.[3]

(b) Fast-growing functions are an old topic of mathematical interest, ever since G. H. Hardy wrote a classic paper about them in 1904. A fast-growing function like our function A was first introduced by Wilhelm Ackermann in 1928 as an example of a computable, but not primitive recursive, function. Now, being recursive, A can be expressed by an L_A wff $A(x, y)$. And – a further feature – PA in fact 'knows' that A is a total function, i.e. knows that for any input x there is a unique output y. In other words, PA can *prove* the truth $\forall x \exists! y A(x, y)$.

However, we can define wilder and wilder fast-growing functions, to the point where they lose that property of being provably-total-according-to-PA. For example, we can define the so-called Goodstein function G; this is an outrageously-hyper-fast-growing recursive function which – like any recursive function – can be canonically expressed in PA by an L_A wff $G(x, y)$. But this time, PA doesn't 'know' that G is a total function, meaning that PA *can't* prove $\forall x \exists! y G(x, y)$ – and since it is sound, PA can't prove the negation of that truth either. (Note though that, unlike the Gödel sentence, *this* new undecidable sentence is not Π_1 but has a greater quantifier complexity.)

It would take us too far afield to define the Goodstein function and explain the proof that it really is a total function and the proof that it isn't provably-total-according-to-PA. Still, the case is worth mentioning for the following reason.

---

[2] Arm-waving more than a bit, any particular p.r. function is computed by a program with 'for' loops nested up to m deep for some fixed m; and that puts a limit on its speed of growth. So, for large enough $n > m$, one of the $f_n(x, x)$ will grow faster.

[3] Arm-waving again, we can numerically code up the stages of a computation of A with input x; we can then do an open-ended search which looks for a number which codes a final computational stage; and then we use a decoding function to extract the value of $A(x)$. So we can re-define $A(x)$ in the form *decode(the least number such that . . .)*, and the minimization in the definition shows we are dealing with a recursive function.

We earlier showed that PA is negation incomplete by coming up with a rather cunningly constructed Gödel sentence which is undecidable. You might very reasonably have wondered: are there arithmetical truths of more ordinary mathematical interest which are expressible in PA but which are likewise undecidable by the theory?

We can now answer: yes, there are. For a start, there are some undecidable sentences which concern outrageously-fast-growing functions.

17.4 Church's Thesis

(a)   The primitive recursive functions are effectively computable. If we add minimization of regular functions to our repertoire of methods for constructing new functions, we get the recursive functions, a more inclusive class of effectively computable functions. So what next? Is there some further way of constructing computable total functions, which takes us beyond the recursive functions?

Perhaps surprisingly, no. Or at least, the following claim is very generally accepted:

Church's Thesis   *The total numerical functions that are effectively computable by some algorithmic routine are just the recursive ones.*

Now, it is debatable whether this is the sort of claim that itself can be *proved* correct, given the imprecision of the informal concept of effective computability: that's why I haven't featured it as a numbered *theorem*. But certainly, every serious attempt to pin down the effectively computable functions in a sharp way does turn out to locate the same class of recursive functions, no less but no more.

For example, Alan Turing famously aimed to analyse the notion of computation in terms of absolutely basic steps (at the level of machine code, if you like). He arrived at the definition of computability by a Turing machine. But we can easily prove that every numerical total function computable by a Turing machine is recursive, and vice versa.

There is a whole family of technical results like this, which we can't explore here. And because of them, Church's Thesis commands almost universal assent. For our purposes, we will accept it too.

(b)   'But not so fast! How can Church's Thesis possibly be acceptable? We used a diagonal argument to show that there are effectively computable functions which aren't primitive recursive. Can't we now run an exactly parallel diagonal argument to show that there are effectively computable functions which aren't recursive?' Good question. But no, we can't 'diagonalize out' of the class of recursive functions. I'll explain why not in §18.4.

(c)   It is worth remarking that Church's Thesis can be used in two different ways, which can call the *interpretive* and the *labour-saving* uses respectively.

The interpretive use involves relying on the Thesis to pass from technical claims about what is or isn't recursive to claims about what is or isn't effectively computable in the intuitive sense. In these cases the Thesis is being used to justify an informal gloss on our technical results.

The labour-saving use involves relying on the Thesis to pass in the opposite direction, from an informal claim to a technical one. In particular, it allows us to jump from a quick-and-dirty informal demonstration that some function is effectively computable to the conclusion that it is recursive. However, any claim about recursiveness argued for in this way *must* also be provable the hard way, without appeal to the Thesis (otherwise we would have located a disconnect between the informal notion of computability and recursiveness, contradicting the Thesis). Hence the labour-saving use of the Thesis must always be inessential.

17.5   Incompleteness theorems again

(a)   Here again is the semantic incompleteness theorem which I initially announced (but, of course, couldn't then prove):

Theorem 1 *Suppose T is an effectively axiomatized formal theory whose language contains the language of basic arithmetic. Then, if T is sound, there will be a true sentence G_T of basic arithmetic such that $T \nvdash G_T$ and $T \nvdash \neg G_T$, so T is negation incomplete.*

By contrast, as noted before, our formal Theorem 44 only applies to appropriate p.r. axiomatized theories.

Now, a p.r. axiomatized theory T is one for which – crucially – the relation Prf_T is primitive recursive. Predictably enough, we will say that a *recursively axiomatized* theory T is one for which – crucially – the relation Prf_T is recursive. Given that L_A can express not just every p.r. function but every recursive function, it is straightforward to check that if Prf_T is recursive then our proof-strategy for Theorem 44 goes through exactly as before. We just have to replace 'primitive recursive' with plain 'recursive' in our earlier proof to give us this strengthened formal result:

Theorem 44* *If T is a sound recursively axiomatized theory whose language contains the language of basic arithmetic, then there will be a true Π_1 sentence G_T such that $T \nvdash G_T$ and $T \nvdash \neg G_T$, so T is negation incomplete.*

That's a straightforward formal result.[4]

Applying Church's Thesis in interpretative mode, however, we can now go on to read the formal 'recursively' here as tantamount to the informal 'effectively'.

---

[4]We can prove this another, less natural, way. By some technical trickery, it can be shown that for any recursively axiomatized theory T there is a perhaps pretty weird but primitive-recursively axiomatized theory T' with the same theorems. Then we can apply our incompleteness theorem for p.r. axiomatized theories to show that T' can't decide $G_{T'}$, and then we can conclude that T can't decide that sentence either since it has the same theorems.

Anyway, see §12.6 again for why a recursively/effectively axiomatized theory which isn't p.r. axiomatized will be a rather strange beast! So the extra strength of this new starred version of our earlier theorem has relatively little impact, except in enabling the tie-in with Theorem 1.

And if we accept that replacement, Theorem 44* gives us, at last, a version of Theorem 1, with the added information that the undecidable sentence is Π_1.

(b)   Very similarly, our initially stated version of the syntactic incompleteness theorem was

Theorem 2 *Suppose T is an effectively axiomatized formal theory whose language contains the language of basic arithmetic. Then, if T is consistent and can prove a certain modest amount of arithmetic (and has an additional property that any sensible formalized arithmetic will share), there will be a sentence G_T of basic arithmetic such that $T \nvdash G_T$ and $T \nvdash \neg G_T$, so T is negation incomplete.*

By contrast, as noted before, our formal Theorem 51 also applies just to appropriate p.r. axiomatized theories. But again, given that Q can capture not only every p.r. function but also every recursive function, it is straightforward to check that, so long as Prf_T is recursive, our proof-strategy for Theorem 51 goes through exactly as before; simply replace 'primitive recursive' with plain 'recursive', giving us another strengthened formal result:

Theorem 51* *If T is a consistent recursively axiomatized theory which contains Q, then there will be a Π_1 sentence G_T such that $T \nvdash G_T$, and if T is ω-consistent, $T \nvdash \neg G_T$. Hence, assuming ω-consistency and so consistency, T is negation incomplete.*

And again, by appeal to Church's Thesis, we can go on to replace 'recursively' here by 'effectively', giving us at last a version of Theorem 2, but now improved by specifying what it takes to 'prove a certain modest amount of arithmetic', and by spelling out that 'an additional property that any sensible formalized arithmetic will share'.

So, as we wanted, we have now joined up the dots, and connected our original informally presented incompleteness theorems with a couple of formal theorems.

18 Decidability, and the Halting Problem

With the more general idea of recursive functions in play, we will now be able to return to and sharpen our earlier Theorem 8, an undecidability result about theories containing enough arithmetic. We will also get a very nice new result about the undecidability of first-order logic. And finally in this chapter we will be able to connect ideas about computability with ideas about incompleteness in an illuminating new way, touching on the famed Halting Problem.

18.1 Q and stronger theories are undecidable

Back in §5.4 we gave a rather elegant proof of

Theorem 8 *No consistent, effectively axiomatized and sufficiently strong formal theory is decidable.*

The informal notion of being 'sufficiently strong' was defined, recall, in terms of capturing all effectively decidable numerical properties. In the light of Church's Thesis, we can trade in this notion for the idea of capturing all recursively decidable properties. And we now know that Q is enough to capture all recursively decidable properties. So here's a new formal analogue of our old theorem:

Theorem 61 *No consistent, recursively axiomatized, formal theory containing* Q *is recursively decidable,*

where, of course, a theory is recursively decidable just if (on a normal Gödel-numbering scheme) the property of numbering a theorem is recursive.

Proof We assume that T is consistent, recursively axiomatized, and contains Q. Now suppose for reductio that T is recursively decidable. That is, suppose that the property $Prov_T$ is recursive (where this is the property which n has iff n Gödel-numbers a T-theorem according to our scheme). If so, like any recursive property, it can be captured by some T-wff $\mathsf{Prv(x)}$.[1]

It is very easily checked that Theorem 55, the Diagonalization Lemma, still applies to recursively axiomatized theories as well as p.r. axiomatized ones. Hence, by the Lemma applied to $\neg\mathsf{Prv}$, there is a T-wff γ such that

---

[1] We needn't take a stance on whether this is, or isn't, the familiar canonical predicate $\mathsf{Prov}_T(\mathsf{x})$.

(i) $T \vdash \gamma \leftrightarrow \neg\mathsf{Prv}(\ulcorner\gamma\urcorner)$.

But by the assumption that Prv captures $Prov_T$, we have in particular

(ii) if $T \vdash \gamma$, i.e. $Prov_T(\ulcorner\gamma\urcorner)$, then $T \vdash \mathsf{Prv}(\ulcorner\gamma\urcorner)$

(iii) if $T \nvdash \gamma$, i.e. not-$Prov_T(\ulcorner\gamma\urcorner)$, then $T \vdash \neg\mathsf{Prv}(\ulcorner\gamma\urcorner)$.

By (i) and (iii), if $T \nvdash \gamma$, then $T \vdash \gamma$. Hence $T \vdash \gamma$. So by (i) and (ii) we have both $T \vdash \mathsf{Prv}(\ulcorner\gamma\urcorner)$ and $T \vdash \neg\mathsf{Prv}(\ulcorner\gamma\urcorner)$ making T inconsistent, contradicting our initial assumption.

So our supposition that T is recursively decidable has to be false. ⊠

Note: by the same argument, while our old friend Prov_T may express the property of numbering a T-theorem, it can't capture it.

18.2 The *Entscheidungsproblem*

Q is such a *very* simple theory that we might quite reasonably have hoped there *would* be some mechanical way of telling which wffs are and which aren't its theorems. But we now know that there isn't. And, in a sense, you can blame the underlying first-order logic.

For, assuming Q's consistency, we get *Church's Theorem* as an immediate corollary of Q's recursive undecidability:

Theorem 62 *The property of being a theorem of first-order logic is not recursively decidable.*

Gödel number the wffs of (your favourite version of) first-order logic; there is no recursive function which, given a code number for a wff, returns a 1/0, yes/no, verdict on the wff's theoremhood.

Proof Since Q has only a finite number of axioms, we can wrap them together into a single conjunction, $\hat{\mathsf{Q}}$. And then, trivially, $\mathsf{Q} \vdash \varphi$ if and only if $\vdash \hat{\mathsf{Q}} \to \varphi$; i.e. we can prove φ inside Q if and only if a certain related conditional is logically provable from no assumptions. So if (1) the relevant first-order logic built into Q were recursively decidable, then (2) we could decide whether the conditional $\hat{\mathsf{Q}} \to \varphi$ is a logical theorem, hence (3) we could recursively decide whether φ is a Q-theorem. But Theorem 61 rules out (3). So (1) must be false – first order logic must be recursively undecidable. ⊠

Hilbert and Ackermann's *Grundzüge der theoretischen Logik* (originally published in 1928) is the first recognizably modern logic textbook, still well worth reading. They posed the *Entscheidungsproblem*, the problem of coming up with an effective method for deciding of an arbitrary sentence of first-order logic whether it is valid or not. Theorem 62 tells us that there is no *recursive* function that takes (the g.n. of) a sentence and returns a verdict on theoremhood.

This doesn't *quite* answer our authors' problem as posed. But Church's Thesis can again be invoked to bridge the gap between our last formal theorem and the following, answering Hilbert and Ackerman:

Theorem 63 *The property of being a theorem of first-order logic is indeed not effectively decidable.*

18.3   Recursive enumerability, theorems, truths

Recall Defn 17 from right back in §5.1. A set Σ is effectively enumerable, we said, iff there is an algorithmic routine which outputs a list of members of the set $s_1, s_2, s_3, \ldots$, repetitions allowed, such that any member of Σ eventually appears on the list if we go on long enough. We then proved Theorem 6, which told us that the theorems of an effectively axiomatized theory T are indeed effectively enumerable.

Using some effective coding scheme, we can trade in items like theorems for numbers, and so turn our attention from effectively enumerable sets in general to effectively enumerable sets of numbers in particular. With that in mind, we can take the following to be the formal analogue of Defn 17:

Defn. 56 *A set of numbers Σ is* recursively enumerable *if and only if it is the range of a (one-place) recursive function.*

Evaluating the recursive function for inputs $0, 1, 2, \ldots$ in turn and we get outputs $n_0, n_1, n_2 \ldots$, thereby effectively enumerating Σ.

Then, corresponding to Theorem 6, we will have

Theorem 64 *The set of (Gödel numbers of) theorems of a recursively axiomatized theory T is recursively enumerable.*

And for our purposes, we can rely on a labour-saving appeal to Church's Thesis in the sense of §17.4 to take us from the old theorem to this new one. But there is of course a technical proof waiting in the wings. Roughly, we keep searching though (codes for) pairs of numbers m, n, to find the next one for which $Prf_T(m, n)$ is true, which is recursively decidable, and then we add n to the list. But we don't need to go into details.

Now for a sharp contrast with the previous theorem, picking out a set which *isn't* recursively enumerable:

Theorem 65 *Let L be a language which includes the language of basic arithmetic: then the set of (Gödel numbers of) its truths is not recursively enumerable.*

Proof Suppose for reductio that there *is* a recursive function f that enumerates the Gödel-numbers for the true L_A sentences (on some normal numbering scheme). A language which includes the language of basic arithmetic suffices to express any recursive function. Hence there will be a formal L-wff $\mathsf{F}(\mathsf{x}, \mathsf{y})$ which expresses that enumerating function f.

And then the formal wff $\exists \mathsf{x}\, \mathsf{F}(\mathsf{x}, \mathsf{y})$ will be satisfied by a number n if and only if n numbers a truth of L. But that contradicts Theorem 60, Tarski's Theorem which tells us that there cannot be such a wff. Hence there can be no enumerating recursive function f. ☒

Putting Theorems 64 and 65 together we get that the set of theorems of a recursively axiomatized theory whose language includes the language of basic arithmetic and set of the truths of that language can't be the same. So assuming the theory is sound, there will be truths it can't prove. Incompleteness, or at any rate the semantic version of the theorem, again!

18.4   The halting problem and incompleteness – very briefly

(a)   In §17.4 I promised to return to the following question: why can't we re-run a diagonal argument on the lines of our argument in §9.5 to show that there are effectively computable functions which aren't recursive?

Recall how the argument went in the case of p.r. functions. The initial claim was that we can effectively generate in turn all the recipes (in some suitable specification language) which define p.r. functions. That way, we get an effective enumeration of p.r. functions, $f_0, f_1, f_2, f_3, \ldots$, repetitions allowed. We then defined the diagonal function $\delta(n) = f_n(n) + 1$ and showed that it was effectively computable but not primitive recursive. Now, in order to run a parallel argument to prove that we can similarly 'diagonalize out' of the class of recursive functions, we'd have to start by showing that we can effectively generate in turn all the recipes (in some suitable specification language) which define recursive functions. But we can't do that. Why? Because there is no effective way of deciding whether a recursive function g is regular; and hence there is no effective way of deciding whether defining $f(x)$ as the least y such that $g(x, y) = 0$ actually gives us a total function; and hence there is no effective way of deciding whether a putative recipe ('starting from these initial functions, repeatedly compose and/or apply recursion/minimalization thus and so") will actually define a recursive function.

(b)   The theorem – not proved here – that there is no effective way of deciding whether a recursive function is regular is just one of a family of closely related results which tell us that there is no effective way of deciding various facts about computations. The senior member of the family – the Halting Problem – is the problem of deciding, given a description of an arbitrary computer program of a particular type and an input, whether the program (i) will halt gracefully giving an output, or (ii) will crash or continue to run forever. Thus Turing proved in 1936 that there is no general effective test for whether a Turing machine will halt gracefully when set running on a given input. In particular, there is a simple diagonal argument – which would take a bit too much setting up to give here – that shows that a Turing machine can't be programmed to decide whether the machine M_e (the e-th in some effective enumeration of possible machines) halts on input e.

And it is not hard to parlay this sort of result into another kind of proof of Gödelian incompleteness. The basic idea is that Turing machines (to continue with that example) are finite objects and we can code up their behaviour using arithmetical codes. So we can construct a formal sentence $H(\bar{e})$ in the language of basic arithmetic which is true if and only if the machine M_e does halt for input e. But now suppose we could effectively enumerate the truths in a language L containing the language of basic arithmetic: then we could set the enumeration going and see which of $H(\bar{e})$ and $\neg H(\bar{e})$ turns up – which gives us an effective way of determining whether M_e halts on input e, which we know we can't have. So there is no effective way of enumerating the truths of L. And if we do the work to avoid a labour-saving appeal to Church's Thesis, we can turn this into another proof of Theorem 65 and hence, as in the last section, of the semantic version of the incompleteness theorem.

(c) An obvious further question arises: in proving incompleteness by thinking about Turing machines halting, can we again drop the semantic assumption that we are dealing with a sound theory of arithmetic and replace it e.g. with the weaker assumption that we are dealing with an ω-consistent theory? We can! And it's rather fun if you like this sort of thing to reprove Theorem 51, our syntactic version of the First Theorem, just from considerations about Turing machines (or we can do it equivalently from considerations about other kinds of programs for computing recursive functions).

Again we can't go into details here.[2] But there is an important moral worth highlighting. Gödelian incompleteness theorems were originally proved by using some apparent trickery with 'self-referring' sentences (which sort-of-say 'I am not provable'). Which can make the theorems seem like oddities. But in fact they can be seen as direct consequences of fundamental limitative results about what computers can and can't do.

---

[2]Enthusiasts who want to know about Turing machines and about the Halting Problem might enjoy looking at *IGT2*, Ch. 43 in particular.

19 The Second Theorem and Hilbert's Programme

It is time at long last for the Second Theorem! As you will see, both its statement and the basic proof-idea are very straightforward; the devil is in the details of a full proof (more about that in the next chapter).

Now, it was easy to see straight away why the First Theorem is so interesting – how odd and unexpected to find that even the truths of basic arithmetic resist being completely pinned down in a nicely axiomatized theory! We will need to do a bit more scene-setting to help bring out the significance of the Second Theorem. So in this chapter we also give a cartoon sketch of some history.

19.1 Defining Con_T

When talking about various axiomatized theories T in earlier chapters, we didn't specify a particular formulation of the deductive system built into T (other than that the resulting theory counts as p.r. axiomatized). The logic may or may not have a built-in *absurdity constant* like the conventional '$\bot$'. Henceforth, then, let's use the absurdity sign in the following way:

Defn. 57 '$\bot$' *is T's built-in absurdity constant if it has one, or else it is an abbreviation for* '$0 = \bar{1}$'.

Assuming T contains Q, T of course proves $0 \neq \bar{1}$. So on either reading of '$\bot$', if T proves $\bot$, it is inconsistent. And conversely, if T has a standard classical (or intuitionistic) logic and T is inconsistent, then on either reading it will prove $\bot$.

Now, assuming again that T contains the language of basic arithmetic and that we have a normal Gödel-numbering scheme in place, then we can express provability-in-T in a canonical way by an open arithmetical T-wff Prov_T, such that $\mathsf{Prov}_T(\ulcorner\varphi\urcorner)$ is true exactly when φ is a T-theorem. Therefore the wff $\neg\mathsf{Prov}_T(\ulcorner\bot\urcorner)$ is true if and only if T *doesn't* prove $\bot$. Hence that wff is true if and only if T is consistent, which motivates

Defn. 58 Con_T *abbreviates* $\neg\mathsf{Prov}_T(\ulcorner\bot\urcorner)$.

For obvious reasons, the arithmetic sentence Con_T is called a (canonical) *consistency sentence* for T. Since Prov_T is Σ_1, Con_T is Π_1.

An example for future use. Take your favourite nicely axiomatized theory of sets – as it might be, ZFC, Zermelo-Fraenkel set theory plus the Axiom of Choice. We can define the language of basic arithmetic in the language of set theory. So ZFC can itself express provability-in-ZFC using a wff of the theory, $\mathsf{Prov_{ZFC}}$, and from that we can form the consistency sentence $\mathsf{Con_{ZFC}}$ which is true iff ZFC is consistent.

And to repeat, this sentence $\mathsf{Con_{ZFC}}$ will be an *arithmetic* sentence (meaning a ZFC version of a sentence of basic arithmetic).[1]

19.2 The Formalized First Theorem

Assume that we are dealing with a theory T which is p.r. axiomatized and contains Q. Then one half of Theorem 51, the syntactic version of the First Theorem, tells us that

(1) If T is consistent, then G_T is not provable in T.

And we now have a natural way of expressing (1) *inside the formal theory T itself*, i.e. by the conditional

(2) $\mathsf{Con}_T \rightarrow \neg\mathsf{Prov}_T(\overline{\ulcorner\mathsf{G}_T\urcorner})$.

Next, let's reflect that, once we have made the cunning move of constructing G_T, the informal reasoning for the First Theorem is in fact *very* elementary. We certainly needed no higher mathematics at all, just relatively straightforward reasoning about arithmetical codings. So we might well expect that if T can reason about arithmetic sufficiently well, *it should itself be able to reflect that elementary reasoning*, and hence itself *prove* (2) – i.e. prove our formalized version of (half of) the First Theorem.

In short, we will hope to have the following key result:

Theorem 66 *If T is p.r. axiomatized and contains enough arithmetic, then* $T \vdash \mathsf{Con}_T \rightarrow \neg\mathsf{Prov}_T(\overline{\ulcorner\mathsf{G}_T\urcorner})$.

Call this the *Formalized First Theorem*.

Of course, that's a skeleton version: we'll want to flesh out what counts as 'enough arithmetic'; but let's leave that for the next chapter. But here's an initial arm-waving thought. Maybe T will need to be stronger that Q, because it will probably need to be able to use induction in generalizing about arithmetized syntax in order to establish (2); but on the other hand, containing PA ought to be more than enough.

---

[1] There are alternatives for defining consistency sentences. Suppose the wff $\mathsf{Contr}(x, y)$ captures the p.r. relation which holds between two numbers when they code for a contradictory pair of sentences, i.e. one codes for some sentence φ and the other for $\neg\varphi$. Then we could define Con_T^* to be short for the sentence $\neg\exists x\exists y(\mathsf{Prov}_T(x) \wedge \mathsf{Prov}_T(y) \wedge \mathsf{Contr}(x, y))$, which says that we can't find two T-provable wffs which contradict each other. But, on modest assumptions, Con_T^* is provably equivalent to Con_T. Another equivalent option would be to use as a consistency sentence the formalization of 'there is a sentence which T doesn't prove'. But we will stick to our standard definition.

19.3   From the Formalized First Theorem to the Second Theorem

Assume for the moment that we *can* establish Theorem 66. In other words, given that T is p.r. axiomatized, and contains enough arithmetic, assume we have

(1) $T \vdash \mathsf{Con}_T \to \neg\mathsf{Prov}_T(\ulcorner\mathsf{G}_T\urcorner)$.

But we know from Theorem 53 that, for a p.r. axiomatized T which contains as little arithmetic as Q,

(2) $T \vdash \mathsf{G}_T \leftrightarrow \neg\mathsf{Prov}_T(\ulcorner\mathsf{G}_T\urcorner)$.

Hence from (1) and (2) we have

(3) $T \vdash \mathsf{Con}_T \to \mathsf{G}_T$.

But we know from the First Theorem (syntactic version) that

(4) If T is consistent, $T \nvdash \mathsf{G}_T$.

And so, from (3) and (4), we get a (somewhat unspecific) version of Gödel's *Second Incompleteness Theorem*.

Theorem 67 *Suppose T is p.r. axiomatized and contains enough arithmetic: then, if T is consistent, $T \nvdash \mathsf{Con}_T$.*

19.4   The impact of the Second Theorem?

Now, if we could have derived Con_T in the theory T, that would of course have been no positive evidence *for* T's consistency. After all, we can derive anything at all in an inconsistent theory! – so in an inconsistent T we could in particular derive Con_T. On the other hand, T's failure to prove Con_T is no evidence *against* T's consistency. We have just found another true Π_1 sentence that T cannot prove, to set alongside G_T.

Hence – you might initially suppose – the non-derivability of a canonical statement of T's consistency inside T does not show us a great deal.

But that's too fast. For consider this obvious corollary of the Second Theorem:

Theorem 68 *Suppose S is a consistent theory, strong enough for the Second Theorem to apply to it, while W is a weaker fragment of S: then $W \nvdash \mathsf{Con}_S$.*

That's because, if S is strong enough for the Second Theorem to apply to it, it can't prove Con_S. Therefore, a fortiori, *part* of S can't prove Con_S either.

So, for example, we *can't* take some strong theory like ZFC as the theory S, use arithmetic coding for talking about its proofs, and then use uncontentious reasoning already available in some weaker, perhaps purely arithmetical, theory, to prove the sentence which codes up the claim that ZFC is consistent. For assuming it *is* consistent, even ZFC itself can't prove $\mathsf{Con}_{\mathsf{ZFC}}$. So assuming the weaker arithmetic theory W is equivalent to a fragment of ZFC, W can't prove $\mathsf{Con}_{\mathsf{ZFC}}$ either.

And *that* is an important result. Why?

19.5   Formalization, finitary reasoning, and Hilbert

Think yourself back to the situation in mathematics near the beginning of the twentieth century.

Classical analysis – the theory of differentiation and integration – has, supposedly, been put on firm foundations. We have, for example, done away with obscure talk about infinitesimals; and we have traded in an intuitive grasp of the continuum of real numbers for the idea of reals defined as 'Dedekind cuts' on the rationals or 'Cauchy sequences' of rationals. And one key idea we've used in our constructions is the idea of a *set* of numbers. In fact, we've been rather free and easy with that idea, allowing ourselves to talk e.g. of arbitrary sets of numbers, even when there is no statable rule for collecting the numbers into the set.

This freedom to allow ourselves to talk of arbitrarily constructed sets is just one aspect of the increasing freedom that mathematicians have allowed themselves over the second half of the nineteenth century. We have loosed ourselves from the assumption that mathematics should be tied to the description of nature: as Morris Kline puts it, "after about 1850, the view that mathematics can introduce and deal with arbitrary concepts and theories that do not have any immediate physical interpretation ... gained acceptance". For example, Cantor could write in 1883 "Mathematics is in its development entirely free and is only bound in the self-evident respect that its concepts must both be consistent with each other and with [other established concepts]".[2]

It is really rather bad news, then, if all this play with freely created concepts, and in particular the fundamental notion of arbitrary sets, actually gets us embroiled in contradiction – as seems to be the case once the set-theoretic paradoxes (like Russell's Paradox) are discovered. What to do?

One response – for example, Russell and Whitehead's in *Principia Mathematica* – is to try to find a secure basis for mathematics in some incontrovertible logical principles. But that project ran into trouble (to be strong enough to do the intended work, *Principia*'s basic principles turned out to be neither obviously logical nor even obviously incontrovertible). So let's consider another sort of response to our foundational troubles. The paradoxes arise within mathematics, and to avoid them (the working mathematican might reasonably think) we just need to do the mathematics more carefully. We don't need to look outside ordinary mathematics for some justification that will guarantee consistency. Perhaps we just need to improve our mathematical practice, in particular by improving the explicitness of our regimentations of mathematical arguments, to reveal the principles we actually use in 'ordinary' mathematics, and to see where the fatal mis-steps must be occurring when we over-stretch these principles in ways that lead to paradox.

How do we improve explicitness? A first step will be to work towards putting the principles that we actually need in mathematics into something approaching

---

[2]Morris Kline, *Mathematical Thought from Ancient to Modern Times* (OUP, 1972) p. 1031. Georg Cantor,*Grundlagen einer allgemeinen Mannigfaltigkeitslehre* (Tuebner 1883) §8.

the ideal form of an effectively axiomatized formal theory. This is what Zermelo aimed to do in axiomatizing set theory: to locate the principles actually needed for the seemingly 'safe' mathematical constructions needed in grounding classical analysis and other familiar mathematical practice. And when the job is done it seems that *these* principles don't in fact allow the familiar reasoning leading to Russell's Paradox or other set-theoretic paradoxes.

So like the logicist camp, the 'do maths better' camp are also keen on regimenting a mathematical theory T into a nice tidy axiomatized format and sharply defining the rules of the game. But the purpose of the axiomatization is quite significantly different. The aim now is not to expose 'foundations', but just to put our set theory or whatever into such a neatly packaged form that – hopefully – we can now show it is indeed consistent and trouble-free.

For note, at this point, a crucial insight of Hilbert's, which we can sum up like this:

> The axiomatic formalization of a mathematical theory T (whether about sets, widgets, or other whatnots), gives us *new* objects (beyond the sets, widgets, or other whatnots) that are *themselves* apt topics for new mathematical investigations – namely the T-wffs and T-proofs that make up the theory!
>
> And, crucially, when we go metatheoretical like this and move from thinking about *sets* (as it might be) to thinking about the syntactic properties of *formalized-theories-about-sets*, we can be moving from considering suites of *infinite* objects to considering suites of *finite* formal objects (the wffs, and the finite sequences of wffs that form proofs). This means that we might then hope to bring to bear, at the metatheoretical level, entirely 'safe', merely *finitary*, reasoning about these suites of finite formal objects in order to prove the consistency of (say) set theory.

Of course, it is a debatable point what exactly constitutes such ultra-'safe' finitary reasoning. But still, it certainly looks as if we will – for instance – need much, much, less than full set theory to in order to reason (not about *sets* but) about a formalized *theory* of sets as a suite of finite syntactic objects. So we might, in particular, hope with Hilbert – in the do-mathematics-better camp – to be able to use a safe uncontentious fragment of finitary mathematics to prove that our wildly infinitary set theory (ZFC, perhaps) is at least syntactically consistent.

At this point you can (I hope!) begin to see the attractions of what's called *Hilbert's Programme* – i.e. the programme of showing various systems of carefully reconstructed infinitary mathematics are contradiction-free by giving consistency proofs using safe, finitary, reasoning about the systems considered as formal objects.

But now enter Gödel, wielding his Second Theorem. He shows that a theory which contains enough arithmetic can't prove its own consistency, let alone the consistency of a stronger theory which includes arithmetic. Hence, for example, we can't use a weak fragment of ZFC – a version of arithmetic, for example –

to prove ZFC consistent by proving Con$_{ZFC}$: not even full-powered ZFC itself can prove Con$_{ZFC}$. Similarly for other strong theories. Which means that the Second Theorem – at least at first blush – sabotages Hilbert's Programme.[3]

Famously, when Hilbert heard of the result at a conference where Gödel first announced it, he was not well pleased ....

---

[3]Fine print. The Second Theorem doesn't rule out all possibility of interesting consistency proofs. For a start, it leaves open the possibility of showing that a theory S is consistent by appeal to a theory S' which is weaker in some respects and stronger in others. Gerhard Gentzen, for example, found such a proof of the consistency of PA. But it would take us too far afield to explore this sort of possibility here (though see *IGT2* §32.4).

20 Proving the Second Incompleteness Theorem

In his original 1931 paper, Gödel states a version of the Second Theorem. But he doesn't spell out a full proof. He simply claims that the reasoning for the First Theorem is so elementary that a strong enough theory must be able formally to replicate the reasoning, so establishing the Formalized First Theorem; and then he notes that this implies the Second Theorem.[1]

The hard work of taking a strong enough T and checking that we get the Formalized First Theorem was first done for a particular case by David Hilbert and Paul Bernays in their *Grundlagen der Mathematik* of 1939. The details of their proof are – the story goes – due to Bernays, who had discussed it with Gödel during a transatlantic voyage.

Now, Hilbert and Bernays helpfully isolated what we can call *derivability conditions* on the predicate Prov_T, conditions whose satisfaction is enough for the Formalized First Theorem to obtain. Later, Martin H. Löb gave a rather neater version of these conditions, giving us the so-called *HBL conditions* which we invoke in this chapter.

20.1 Sharpening the Second Theorem

Recall: $\mathsf{Prov}_T(\mathsf{x})$ abbreviates $\exists \mathsf{v}\, \mathsf{Prf}_T(\mathsf{v}, \mathsf{x})$; and Prf_T is a Σ_1 expression. So arguing about provability inside T will involve establishing some general claims involving Σ_1 expressions. And how do we prove general arithmetical claims? Using induction is the default method.

It therefore looks to be a good bet that we will get $T \vdash \mathsf{Con}_T \to \neg\mathsf{Prov}_T(\ulcorner \mathsf{G}_T \urcorner)$ if T at least has Σ_1-induction – meaning that T's axioms include (the universal closures of) all instances of the first-order Induction Schema where the induction predicate φ is no more complex than Σ_1.

In §8.6 we introduced $\mathsf{I}\Sigma_1$ as the standard label for the result of adding that amount of induction to Q. So it is a plausible conjecture that we can show the following, a fleshed-out version of our skeletal Theorem 66:

---

[1] I am told, though, that Gödel's shorthand notebooks at the time suggest that he in fact hadn't then worked out any detailed proof of the first step.

Theorem 66* *If T is p.r. axiomatized and contains* $\mathsf{I}\Sigma_1$, *then* $T \vdash \mathsf{Con}_T \to \neg\mathsf{Prov}_T(\ulcorner\mathsf{G}_T\urcorner)$.

And if that is right, by the argument of §19.3 we get an improved version of Theorem 67, a sharper Second Theorem:

Theorem 67* *If T is consistent, p.r. axiomatized and contains* $\mathsf{I}\Sigma_1$, *then* $T \nvdash \mathsf{Con}_T$.

20.2   The box notation

To improve readability, we introduce some standard notation. We will henceforth abbreviate $\mathsf{Prov}_T(\ulcorner\varphi\urcorner)$ simply by $\Box_T\varphi$ – you can read that as 'φ is provable (in T)'. This is perhaps a bit naughty, as the new notation hides away the Gödel-numbering and the use of standard numerals for the codes. But the notation is standard and is safe enough if we keep our wits about us.[2]

So in particular, $\neg\mathsf{Prov}_T(\ulcorner\mathsf{G}_T\urcorner)$ can be abbreviated $\neg\Box_T\mathsf{G}_T$. Thus in our new notation, the Formalized First Theorem is $\mathsf{Con}_T \to \neg\Box_T\mathsf{G}_T$. Moreover, Con_T can now alternatively be rendered as $\neg\Box_T\bot$.

20.3   Proving the Formalized First Theorem

First a standard definition: we will say (dropping subscripts)

Defn. 59 *The HBL derivability conditions* hold in T *if and only if, for any T-sentences φ, ψ,*

   C1. If $T \vdash \varphi$, then $T \vdash \Box\varphi$;

   C2. $T \vdash \Box(\varphi \to \psi) \to (\Box\varphi \to \Box\psi)$;

   C3. $T \vdash \Box\varphi \to \Box\Box\varphi$.

Here, (C1) tells us that if T can prove φ, then (via the relevant Gödel-numbering which enables T to code up claims about provability-in-T), T as it were 'knows' that φ is provable. (C2) tells us that T 'knows' about modus ponens; in other words, if T can show $(\varphi \to \psi)$ is provable, and can show φ is provable, then T can show that ψ is provable too. And (C3) tells us that if T 'knows' it can prove φ, it can code up that proof so it can prove that φ is provable. We might reasonably hope that these conditions will hold for a strong enough theory T.

And we can indeed show that

Theorem 69 *If T is p.r. axiomatized and contains* $\mathsf{I}\Sigma_1$, *then the derivability conditions hold for T.*

---

[2]If you are familiar with modal logic, then you will immediately recognize the symbol conventionally used for a necessity operator. And the parallels and differences between ' "$1 + 1 = 2$" is provable (in T)' and 'It is necessarily true that $1 + 1 = 2$' are highly suggestive. These parallels and differences are the topic of 'provability logic', the subject of a contemporary classic, George Boolos's *The Logic of Provability* (CUP, 1993).

However, demonstrating this is a seriously tedious task, and I certainly *don't* propose to hack through the annoying technical details. In this book we will allow ourselves to take this technical Theorem 69 as given.

We can also show, considerably more easily, that

Theorem 70 *If T is p.r. axiomatized, contains Q, and the derivability conditions hold for T, then the Formalized First Theorem holds.*

Proof This is just a mildly fun exercise in box-juggling, with the target of showing $T \vdash \mathsf{Con}_T \to \neg\Box_T \mathsf{G}_T$.

First, since T is p.r. axiomatized and contains Q, Theorem 53 holds. So, in our new symbolism and dropping subscripts, we have $T \vdash \mathsf{G} \leftrightarrow \neg\Box\mathsf{G}$.

Second, since T contains Q and standard logic, we have

$$T \vdash \neg\varphi \to (\varphi \to \bot).$$

(whichever way we read the absurdity constant – see Defn. 57). Given this and the derivability condition (C1), we get

$$T \vdash \Box(\neg\varphi \to (\varphi \to \bot)).$$

So given the derivability condition (C2) and using modus ponens, it follows that for any φ

(A) $T \vdash \Box\neg\varphi \to \Box(\varphi \to \bot)$.

We now argue as follows:

(1)	$T \vdash \mathsf{G} \to \neg\Box\mathsf{G}$	Half of Theorem 53
(2)	$T \vdash \Box(\mathsf{G} \to \neg\Box\mathsf{G})$	From 1, given C1
(3)	$T \vdash \Box\mathsf{G} \to \Box\neg\Box\mathsf{G}$	From 2, using C2
(4)	$T \vdash \Box\neg\Box\mathsf{G} \to \Box(\Box\mathsf{G} \to \bot)$	Instance of A
(5)	$T \vdash \Box\mathsf{G} \to \Box(\Box\mathsf{G} \to \bot)$	From 3 and 4
(6)	$T \vdash \Box\mathsf{G} \to (\Box\Box\mathsf{G} \to \Box\bot)$	From 5, using C2 and logic
(7)	$T \vdash \Box\mathsf{G} \to \Box\Box\mathsf{G}$	Instance of C3
(8)	$T \vdash \Box\mathsf{G} \to \Box\bot$	From 6 and 7
(9)	$T \vdash \neg\Box\bot \to \neg\Box\mathsf{G}$	Contraposing
(10)	$T \vdash \mathsf{Con} \to \neg\Box\mathsf{G}$	Definition of Con ⊠

Now we put the last two theorems together. If T is p.r. axiomatized and contains $\mathsf{I\Sigma}_1$, the derivability conditions will hold (by Theorem 69) and it contains Q. Hence (by Theorem 70) if T is p.r. axiomatized and contains $\mathsf{I\Sigma}_1$, it proves the Formalized First Theorem. Which gives us our target Theorem 67*.

20.4   The equivalence of Con_T with G_T

(a)   We'll assume throughout this section that T is a p.r. axiomatized theory such that the derivability conditions hold. And first, we will show that in this case, G_T and Con_T are provably equivalent in T.

We have just shown in the previous Theorem that $T \vdash \mathsf{Con} \to \neg\Box\mathsf{G}$. We can now prove the converse, $T \vdash \neg\Box\mathsf{G} \to \mathsf{Con}$. Indeed, we have a more general result:

Theorem 71 *For any sentence* φ, $T \vdash \neg\Box\varphi \rightarrow$ Con.

Proof We argue as follows:

(1)	$T \vdash \bot \rightarrow \varphi$	Logic!
(2)	$T \vdash \Box(\bot \rightarrow \varphi)$	From 1, given C1
(3)	$T \vdash \Box\bot \rightarrow \Box\varphi$	From 2, given C2
(4)	$T \vdash \neg\Box\varphi \rightarrow \neg\Box\bot$	Contraposing
(5)	$T \vdash \neg\Box\varphi \rightarrow$ Con	Definition of Con ⊠

This little theorem has a rather remarkable corollary. Since T can't prove Con, the theorem tells us that T doesn't entail $\neg\Box\varphi$ for any φ at all. Hence T doesn't ever 'know' that it can't prove φ, even when it can't! In sum, suppose that T satisfies the now familiar conditions: by (C1), T knows all about what it *can* prove; but we have just shown that it knows nothing about what it *can't* prove.

(b)   As a particular instance of the last theorem, we have $T \vdash \neg\Box G \rightarrow$ Con. So putting that together with Theorem 70, we have $T \vdash$ Con $\leftrightarrow \neg\Box G$. And now combine *that* with Theorem 53 which tells us that $T \vdash G \leftrightarrow \neg\Box G$, and lo and behold we've shown

Theorem 72 *If T is p.r. axiomatized and contains* $\mathsf{I}\Sigma_1$, *then* $T \vdash$ Con $\leftrightarrow$ G.

This means that, not only do we have $T \nvdash$ Con, we also have (assuming in addition that T is ω-consistent) $T \nvdash \neg$Con. In other words, Con is formally undecidable by T.

(c)   To continue: we next prove

Theorem 73 $T \vdash$ Con $\leftrightarrow \neg\Box$Con.

Proof The direction from right to left is an instance of Theorem 71. For the other direction, note that

(1)	$T \vdash$ Con $\rightarrow$ G	Already proved
(2)	$T \vdash \Box($Con $\rightarrow$ G$)$	From 1, given C1
(3)	$T \vdash \Box$Con $\rightarrow \Box$G	From 2, given C2
(4)	$T \vdash \neg\Box$G $\rightarrow \neg\Box$Con	Contraposing
(5)	$T \vdash$ Con $\rightarrow \neg\Box$Con	Using Thm. 70 ⊠

So: this shows that Con (like G) is also a fixed point of the negated provability predicate (see again Defn. 48 and §14.2).

As we have noted before (§14.3, fn. 1), some authors refer to any fixed point of the negated provability predicate as a Gödel sentence. Fine. That's one way of using the jargon. But if you adopt the broad usage, you must be careful with your informal commentary. For example, not all Gödel sentences in the broad sense indirectly 'say' *I am unprovable*: Con is a case in point – it is *not* a wff which is more or less directly 'about' itself. That observation should scotch any lingering suspicion that undecidable sentences provided by broadly Gödelian arguments are all tainted by potentially paradoxical self-reference.

20.5 Löb's Theorem

Now that we have the derivability conditions in play, let's state and prove a theorem from 1955, due to Martin Löb:

Theorem 74 *If T is p.r. axiomatized and contains $I\Sigma_1$ (so the derivability conditions hold), then for any T-wff φ, if $T \vdash \Box\varphi \to \varphi$ then $T \vdash \varphi$.*

Proof More box juggling! Assume that, for a given φ,

(1) $T \vdash \Box\varphi \to \varphi$.

And next consider the wff $\mathsf{Prov}(\mathsf{x}) \to \varphi$. Given our assumptions about T, the Diagonalization Lemma applies. Hence for some δ, T proves $\delta \leftrightarrow (\mathsf{Prov}(\ulcorner\delta\urcorner) \to \varphi)$. Or, using our box notation,

(2) $T \vdash \delta \leftrightarrow (\Box\delta \to \varphi)$.

And now we can continue as follows:

(3)	$T \vdash \delta \to (\Box\delta \to \varphi)$	From 2
(4)	$T \vdash \Box(\delta \to (\Box\delta \to \varphi))$	From 3, by C1
(5)	$T \vdash \Box\delta \to \Box(\Box\delta \to \varphi)$	From 4, by C2
(6)	$T \vdash \Box\delta \to (\Box\Box\delta \to \Box\varphi)$	From 5, by C2
(7)	$T \vdash \Box\delta \to \Box\Box\delta$	By C3
(8)	$T \vdash \Box\delta \to \Box\varphi$	From 6 and 7
(9)	$T \vdash \Box\delta \to \varphi$	From 1 and 8
(10)	$T \vdash \delta$	From 2 and 9
(11)	$T \vdash \Box\delta$	From 10, by C1
(12)	$T \vdash \varphi$	From 9 and 11 ⊠

This is perhaps rather a surprise. We might have expected that a theory which has a well-constructed provability-predicate should 'think' that, quite generally, if it can prove φ then indeed φ – i.e. we might have expected that in general $T \vdash \Box\varphi \to \varphi$. But not so. A respectable theory T can only prove this if in fact it can *already* show that φ.

By the way, Löb's Theorem answers a question raised by Leon Henkin. We know from the Diagonalization Lemma applied to the *negated* wff $\mathsf{Prov}(\mathsf{x})$ that there is a sentence G such that $T \vdash \mathsf{G} \leftrightarrow \neg\mathsf{Prov}(\ulcorner\mathsf{G}\urcorner)$, and that a consistent T can't prove G. Similarly, we can apply the Lemma to the *unnegated* wff $\mathsf{Prov}(\mathsf{x})$ to show that there is a sentence H such that $T \vdash \mathsf{H} \leftrightarrow \mathsf{Prov}(\ulcorner\mathsf{H}\urcorner)$ – so it is as if H says 'I am provable'. Henkin asked: *is* H provable? We now know that it is. For by hypothesis, $T \vdash \mathsf{Prov}(\ulcorner\mathsf{H}\urcorner) \to \mathsf{H}$, i.e. $T \vdash \Box\mathsf{H} \to \mathsf{H}$; so $T \vdash \mathsf{H}$ by Löb's Theorem.

Löb's Theorem also immediately entails the Second Theorem again. For assume the conditions for Löb's Theorem apply. Then as a special case we get that if $T \vdash \Box\bot \to \bot$ then $T \vdash \bot$. Therefore, if $T \nvdash \bot$, so T is consistent, then $T \nvdash \Box\bot \to \bot$, hence $T \nvdash \neg\Box\bot$, hence (just by definition of Con), $T \nvdash \mathsf{Con}$.

There's a converse too, i.e. the Second Theorem fairly quickly entails Löb's Theorem.[3] So the two theorems come to much the same.

And it is worth noting how the proof of Löb's Theorem from the biconditional at line (2) to the conclusion at line (12) is very reminiscent of the argument for Curry's Paradox which we met in §16.3; and we can with a bit of massaging make the arguments run more closely in parallel. So the situation is this. What Gödel saw in proving the first theorem, in the broadest terms, is that when we move from talking about truth to talking about provability, thinking about Liar-style sentences can lead not to paradox but to the First Incompleteness Theorem. Much later, Löb spotted that, similarly, when we move from talking about truth to talking about provability, Curry-paradox style reasoning again leads not to paradox but to the Second Theorem.

---

[3]For the proof idea due to Kripke, see *IGT2*, §34.5.

21 Complications

There are some intriguing complications with the second theorem. For a start there are consistent theories that seemingly 'prove their own inconsistency'; and there are, despite the Second Theorem, senses in which a theory can 'prove its own consistency'. There is, relatedly, an important but easy-to-miss difference between what is involved in proving the first theorem and what the second theorem requires.

21.1 Consistent theories that 'prove their own inconsistency'

An ω-*consistent* (and so consistent) p.r. axiomatized theory T which contains a little arithmetic can't prove $\neg\mathsf{Con}_T$, as we've seen. By contrast, as we'll show next, a consistent but ω-*inconsistent* T can have $\neg\mathsf{Con}_T$ as a theorem. (Being ω-inconsistent, T won't be sound; that's how it can have a false theorem!)

The proof is actually straightforward, once we note a simple lemma. Suppose W and S are two p.r. axiomatized theories, weaker and stronger, which share a deductive logic; and suppose every axiom of the weaker theory W is also an axiom of the stronger theory S. It is then a trivial logical truth that, if the stronger theory S is consistent, then the weaker theory W must be consistent too. Assuming that S contains Q, it will be able to formally prove the arithmetical claim that encodes this fact:

Theorem 75 *Under the given conditions, with theory S extending theory W,* $S \vdash \mathsf{Con}_S \to \mathsf{Con}_W$.

So now take our simpler theory W to be PA. Take the stronger theory S to be PA augmented by the extra axiom $\neg\mathsf{G}_{\mathsf{PA}}$. We have met this richer theory briefly before and, by Theorem 50, S is consistent but ω-inconsistent. Since, quite trivially $S \vdash \neg\mathsf{G}_{\mathsf{PA}}$, and PA (and hence S) proves $\mathsf{G}_{\mathsf{PA}} \leftrightarrow \mathsf{Con}_{\mathsf{PA}}$, $S \vdash \neg\mathsf{Con}_{\mathsf{PA}}$. So, using our last theorem and modus tollens, $S \vdash \neg\mathsf{Con}_S$.

Summing that up,

Theorem 76 *Assuming PA is consistent, the theory $S = \mathsf{PA}+\neg\mathsf{G}_{\mathsf{PA}}$ is a consistent theory which 'proves' itself inconsistent (i.e. we can derive in it the negation of its own consistency sentence).*

And since $S \vdash \neg\mathsf{Con}_S$,

136

Theorem 77 *There can be a* consistent *theory S such that the theory $S + \mathsf{Con}_S$ is* inconsistent.

What are we to make of these apparent absurdities? Well, giving the language of S, i.e. $\mathsf{PA} + \neg \mathsf{G_{PA}}$, its standard built-in arithmetical interpretation, the theory is unsound (since the Gödel sentence $\mathsf{G_{PA}}$ is true). So we shouldn't trust what S says about anything, especially about its own inconsistency when we derive $\neg \mathsf{Con}_S$ from it. S doesn't really *prove* (in the ordinary sense of establish-as-true) its own inconsistency, since we don't accept the theory as correct on the standard interpretation! That's why we used scare quotes in stating Theorem 76.

21.2   There are provable consistency sentences

We now turn to discuss a way in which, despite the Second Theorem, consistent theories *can* prove sentences which express their own consistency. As you will see, this involves a simple trick which deprives the result of much intrinsic interest: however, it does point up an important moral which we will draw out in the final section.

Assume that we are dealing with a theory satisfying the usual conditions, with a Gödel-numbering scheme in place. Roughly what we are going to do is tweak the predicate Prf_T which canonically captures Prf_T to get another predicate CPrf_T which still expresses and captures Prf_T (assuming T is consistent), form a new provability predicate CProv_T from *that*, and use this to define a new consistency sentence CCon. We then show that T can (quite trivially) prove CCon_T – in fact, even Q can prove that, for any T.

(a)   First step. Suppose we put

Defn. 60 $\mathsf{CPrf}_T(\mathsf{x}, \mathsf{y}) =_{\text{def}} \mathsf{Prf}_T(\mathsf{x}, \mathsf{y}) \wedge (\forall \mathsf{w} \leq \mathsf{x}) \neg \mathsf{Prf}_T(\mathsf{w}, \overline{\ulcorner \bot \urcorner})$.

Assuming T is consistent, CPrf_T expresses Prf_T.

Why? If T is consistent and $Prf_T(m, n)$, then (i) $\mathsf{Prf}_T(\overline{m}, \overline{n})$ will be true, and (ii) given T's consistency, no number smaller or equal to m will code for a proof of absurdity, so $(\forall \mathsf{w} \leq \overline{m}) \neg \mathsf{Prf}_T(\mathsf{w}, \overline{\ulcorner \bot \urcorner})$ will be true too. Hence $\mathsf{CPrf}_T(\overline{m}, \overline{n})$ will be true.

While if $Prf_T(m, n)$ fails, then $\mathsf{Prf}_T(\overline{m}, \overline{n})$ is false, and trivially $\mathsf{CPrf}_T(\overline{m}, \overline{n})$ will be false too.

It can also be shown that CPrf_T captures Prf_T. We just use an argument parallel to the one in §15.1 that shows that the very similar Rosserized predicate RPrf_T also captures Prf_T.

(b)   Next, predictably, we define

Defn. 61 $\mathsf{CProv}_T(\mathsf{x}) =_{\text{def}} \exists \mathsf{v} \mathsf{CPrf}_T(\mathsf{v}, \mathsf{x})$.

The code number for a wff φ satisfies this revised provability predicate $\mathsf{CProv}_T(\mathsf{x})$ just if there is a T-proof of φ, while there is no T-proof of absurdity which comes earlier than that proof of φ (when proofs are ordered by their code numbers).

(c)   Now, there might in fact seem to be some real motivation for being interested in such a tweaked provability predicate. For consider the following line of thought.

When trying to establish an as-yet-unproved conjecture, mathematicians will use any tools to hand, bringing to bear whatever background assumptions that they are prepared to accept in the context. The more improvisatory the approach, the less well-attested the assumptions, then the greater the risk of lurking inconsistencies emerging, requiring our working assumptions to be revised. We should therefore – in an ideal world – keep a running check on whether apparent new results cohere with secure background knowledge. Only a derivation which passes the coherence test has a chance of being accepted as a kosher *proof*. So how might we best reflect something of the idea that a genuine proof should be (as we might put it) *consistency-minded*, i.e. should come with a certificate of consistency with what's gone before?

We might reasonably suggest as a minimum: there is a consistency-minded proof of φ in the axiomatized formal system T only if (i) there is an ordinary T-derivation of φ with super g.n. m, while (ii) there isn't already a T-derivation of absurdity with a code number less than m. That's the idea which is formally reflected in the provability predicate CProv.

(d)   So let's now define a corresponding consistency sentence

Defn. 62 $\text{CCon}_T =_{\text{def}} \neg\text{CProv}_T(\ulcorner\bot\urcorner)$.

Since CProv_T expresses a provability relation, CCon_T (at first sight) 'says' that the relevant T is consistent. But it is immediate that so long as T contains Q,

Theorem 78 $T \vdash \text{CCon}_T$.

Proof Logic gives us $\neg(\text{Prf}_T(\mathsf{a},\ulcorner\bot\urcorner)\wedge\neg\text{Prf}_T(\mathsf{a},\ulcorner\bot\urcorner))$, where a is a dummy name. That entails $\neg(\text{Prf}_T(\mathsf{a},\ulcorner\bot\urcorner) \wedge (\forall\mathsf{v}\leq\mathsf{a})\neg\text{Prf}_T(\mathsf{v},\ulcorner\bot\urcorner))$, since Q can handle the bounded quantifier. Apply universal quantifier introduction to infer CCon_T.   ⊠

(e)   What can we learn about the consistency of a theory T from last theorem? Predictably, nothing at all!

Take T to be any theory of interest, e.g. Zermelo-Fraenkel set theory. Now, if T is indeed consistent, CProv_T indeed expresses provability-in-T, and CCon_T 'says' that T is consistent. But if T isn't consistent, then CProv_T doesn't reliably express that proof relation (suppose φ is provable but every proof of it has a larger g.n. than a proof of absurdity; then we won't have $\text{CProv}_T(\ulcorner\varphi\urcorner)$). So if T isn't consistent, CCon_T can't necessarily be read as expressing consistency. *But this means that we already need to know whether T is consistent before we can interpret what we've proved in deriving CCon_T. Therefore the proof can't tell us whether T is consistent!*[1]

---

[1] It might help to think about this related example. Take a theory T. Let S be T's largest consistent sub-theory (we can make that idea respectable in various ways). Suppose T is in fact consistent. Then S is none other than T! Does this mean, in this case, that the trivial observation that S is consistent is a proof that T is consistent?

21.3   The 'intensionality' of the Second Theorem

The provability of Theorem 78 tells us nothing about the consistency of the relevant theory T. But it *does* tell us something important about what it takes to prove the Second Theorem.

Recall, using the box shorthand, we showed that if the HBL derivability conditions hold for the provability predicate Prov_T, we can't derive in T the consistency sentence Con_T (i.e. we can't derive $\neg\mathsf{Prov}_T(\ulcorner\bot\urcorner)$). Since we can derive in T the sentence CCon_T (i.e. we can derive $\neg\mathsf{CProv}_T(\ulcorner\bot\urcorner)$), the derivability conditions can't all apply to CProv_T. What makes the difference?

In fact, in showing that the derivability conditions do hold for Prov_T, we have to rely on the fact that Prov_T is defined in terms of a predicate expression Prf_T which *canonically* captures Prf_T. Putting it roughly, we need $\mathsf{Prf}_T(\overline{m}, \overline{n})$ to reflect the details of what it takes for m to code a proof of the wff with g.n. n, and no more. Putting it even more roughly, we need Prf_T to have the right intended *meaning*. For if we doctor Prf_T to get another wff CPrf_T which still captures Prf_T but in a 'noisy' way, then – as we saw – the corresponding CProv_T need not satisfy the derivability conditions.

So we have the following important difference between the First and the Second Theorem. Suppose T is p.r. axiomatized and contains Q:

(1) Take *any* old wff Prf_T^* which captures Prf, in however 'noisy' a way. Form the corresponding proof predicate Prov_T^*. Take a fixed point γ for $\neg\mathsf{Prov}_T^*(\mathsf{x})$. Then we can show, just as before, that $T \nvdash \gamma$, assuming T is consistent, and $T \nvdash \neg\gamma$ assuming ω-consistency as well. (The proof depends just on the fact that Prf_T^* successfully captures Prf_T, but not at all on *how* it does the job.)

(2) By contrast, if we want to prove $T \nvdash \mathsf{Con}_T$ (assuming T is consistent and has a smidgin of induction), we have to be a lot more picky. We need to start with a wff Prf_T which does 'have the right meaning', i.e. which *canonically* captures Prf_T, and then form the canonical consistency sentence Con_T from *that*. It we capture Prf_T in the wrong way, using a doctored Prf_T^*, then the corresponding Con_T^* may indeed be provable after all.

Solomon Feferman describes results which depend on the open wff Prf_T "more fully express[ing] the notion involved" as 'intensional'.[2] The label, associated with issues of meaning, carries baggage which we needn't worry about. The important thing is the easy-to-miss distinction Feferman wants to emphasize, between what it takes to prove the Second Theorem as opposed to the First.

---

[2]In his 'Arithmetization of Metamathematics in a General Setting', *Fundamenta Mathematicae* (1960), p. 35.

Appendix: Kripke on diagonalization

Our proof of the standard form of the Diagonalization Lemma was not difficult. But it relied on a not-immediately-obvious initial construction. So I'll consider here a variant due to Saul Kripke, which he claims to be more natural.[1]

Suppose throughout what follows that T is the usual kind of theory, p.r. axiomatized and containing Q (and so incorporating the language of basic arithmetic). And assume we have a Gödel-numbering scheme in place.

We can effectively list T's open wffs with the free variable x in the usual way $\varphi_0(x), \varphi_1(x), \varphi_2(x), \varphi_3(x), \ldots$. Now add to T's language a corresponding infinite sequence of *new* constants $c_0, c_1, c_2, c_3, \ldots$, and expand our Gödel-numbering scheme to give code numbers to each of the new constants.[2] Then:

Defn. 63 *The* Kripke extension T^K *of the theory* T *whose open wffs with free variable* x *are* $\varphi_i(x)$ *is the result of adding all the new constants* c_i *to* T's *language, and adding to* T's *axioms all the sentences* $c_i = \ulcorner \varphi_i(c_i) \urcorner$.

The idea, then, is to introduce constants c_i governed by axioms which ensure that a sentence $\varphi_i(c_i)$ is – via the coding – more directly 'about' itself, without the usual diagonalization trickery.

We need to read into the record six quick and easy observations:

(O1) T^K is still p.r. axiomatized. We can still effectively determine (without open-ended searches) whether a given wff is an axiom, since there is a simple test whether a wff has the form of one of the new axioms.

(O2) Any wff without c-constants which is provable in T^K is already provable in T. If we add new constants which are in effect just shorthand for certain standard numerals (different constants, different numerals), then the only new wffs we can prove will be wffs which feature those constants.

(O3) Any wff of T^K with some c-constant(s) is provably-in-T^K equivalent to a wff without c-constants. Just use the c-axioms and Leibniz's Law to replace a constant with a standard numeral.

---

[1] Kripke's construction has long been folklore but was only published in 2020, in his 'Gödel's Theorem and direct self-reference' (*arXiv* 2010.11979).

[2] This harmlessly assumes that our scheme is like the one introduced in §11.1 in having an infinite sequence of as-yet-unused code numbers available to code the new constants. We can always arrange this to be so, if only by the low trick of doubling all the Gödel numbers in our initially assumed scheme, leaving us with all the odd numbers to code the c_i!

(O4) For any wff α of T's language, if $T^K \nvdash \alpha$, then $T \nvdash \alpha$. Trivially, if the theory with extra axioms can't prove α, then the original theory can't.

(O5) If T is consistent, so is T^K. Again, adding distinct new constants for distinct numerals can't make a consistent theory inconsistent.

(O6) If T is ω-consistent, so is T^K. We prove the contrapositive. Suppose T^K is ω-inconsistent, so there is an open wff $\psi(x)$, such that $T^K \vdash \exists x \psi(x)$ yet for each number n we have $T^K \vdash \neg\psi(\bar{n})$. Apply (O3) to ψ if necessary to get a provably equivalent ψ' without any of the new constants, and then $T^K \vdash \exists x \psi'(x)$ while for each number n we have $T^K \vdash \neg\psi'(\bar{n})$. But since these wffs in ψ' are free of c-constants, we can apply (O2) to get $T \vdash \exists x \psi'(x)$ while for each number n we have $T \vdash \neg\psi'(\bar{n})$, which makes the original T ω-inconsistent.

We now have the following variant of (the second part) of the original Lemma:

Theorem 79 *If φ is a one-place open sentence of T's language, then there is a sentence δ of T's Kripke extension T^K such that $T^K \vdash \delta \leftrightarrow \varphi(\ulcorner \delta \urcorner)$.*

Proof Trivially, φ will be some φ_i. By the relevant c-axiom, $T^K \vdash \varphi_i(c_i) \leftrightarrow \varphi_i(\ulcorner \varphi_i(c_i) \urcorner)$. So just put $\delta = \varphi_i(c_i)$ and we are done. ⊠

That was easy! Now take the case where $\varphi(x)$ is $\neg\mathsf{Prov}_{T^K}(x)$, the negation of the canonical provability predicate for T^K. This is formed in the (c_i-free) language of basic arithmetic, i.e. is a one-place predicate already available in T. So we can invoke our theorem to show that there is in particular a fixed point γ of the language of T^K such that $T^K \vdash \gamma \leftrightarrow \neg\mathsf{Prov}_{T^K}(\ulcorner \gamma \urcorner)$.

And now we can apply Theorem 54. Since we have a fixed point γ for $\neg\mathsf{Prov}_{T^K}$, we know that (i) if T^K is consistent, $T^K \nvdash \gamma$. And (ii) if T^K is ω-consistent, $T^K \nvdash \neg\gamma$.

Fine. But this doesn't quite get us back to the desired incompleteness theorem for our original theory T. For a start, the fixed point γ (on our current construction) involves one of those new constants, so γ won't belong to the language of T. So we now need to use our observations above.

By (O3), we know that there is a γ' without new constants which is provably equivalent to γ, so (i') if T^K is consistent, $T^K \nvdash \gamma'$; and (ii') if T^K is ω-consistent, $T^K \nvdash \neg\gamma'$. So then applying (O5) and (O4), we get (i'') if T is consistent, so is T^K, so $T^K \nvdash \gamma'$, and hence $T \nvdash \gamma'$. And applying (O6) and (O4) we get (ii'') if T is ω-consistent so is T^K, so $T^K \nvdash \neg\gamma'$, hence $T \nvdash \neg\gamma'$. Which gives us the First Incompleteness Theorem for T again.

In summary, then. To get incompleteness from a Diagonalization Lemma, whether our way or Kripke's, you rely on the key Theorem 54. With our original Lemma, you get to apply Theorem 54 'neat'. If you use Kripke's construction to arrive at an easier variant Lemma (adding constants stipulated to do the necessary work), then this time you have to take a bit of care to massage away the constants after you've applied Theorem 54. So you pays your money and you takes your choice!

Further reading

And here we will end (as did the lectures these notes were originally written to accompany). But let me mention three books which take the story further, and also give a pointer to a resource which suggests additional references.

First, though, a general remark. We have seen that Gödel's 1931 proof of his incompleteness theorem uses facts about primitive recursive functions: and these functions are a subclass, but only a subclass, of the effectively computable numerical functions. A more general treatment of the effectively computable functions (capturing *all* of them, if we are to believe Church's Thesis) was developed in the mid–1930s, and this in turn throws more light on the incompleteness phenomenon, as briefly indicated in §§17.5 and 18.4. So there's a choice to be made. Do you look at things in roughly the historical order, first encountering just the primitive recursive functions and learning how to prove initial versions of Gödel's incompleteness theorem before moving on to look at the general treatment of computable functions? Or do you do some of the general theory of computation first, turning to the incompleteness theorems later?

In this book we have very largely taken the first route. You can explore this further in *IGT2*, i.e.

> Peter Smith, *An Introduction to Gödel's Theorems* (CUP, 2nd edition 2013).

A corrected version is freely downloadable from logicmatters.net/igt, where there are also supplementary materials of various kinds.

Alternatively, you can take the second route by following

> Richard Epstein and Walter Carnielli, *Computability: Computable Functions, Logic, and the Foundations of Mathematics* (Wadsworth 2nd edn. 2000: Advanced Reasoning Forum 3rd edn. 2008)

This explores general computability theory first. It is a very good introductory text on the standard basics, particularly clearly and attractively done, with a lot of interesting and illuminating historical information too in Epstein's 28 page timeline on 'Computability and Undecidability' at the end of the book.

Those two books should be very accessible to those without much mathematical background, but even experienced mathematicians should appreciate the careful introductory orientation which they provide. Then, going up just half a step in mathematical sophistication, we arrive at a really beautiful book:

George Boolos and Richard Jeffrey, *Computability and Logic* (CUP 3rd edn. 1990).

This is a modern classic, wonderfully lucid and engaging. There are in fact later editions – heavily revised and considerably expanded – with John Burgess as a third author. But I know that I am not the only reader to think that the later versions (excellent though they are) do lose something of the original book's famed elegance and individuality. Still, whichever edition comes to hand, do read it! – you will learn a great deal in an enjoyable way.

Then, for many more references at various levels, you can see the relevant sections of the *Beginning Mathematical Logic* Study Guide, downloadable from logicmatters.net/tyl.

Index of definitions

CPSIA information can be obtained
at www.ICGtesting.com
Printed in the USA
BVHW011343231122
652621BV00004B/66